TELEPEN
AF598436
WITHDRAWN

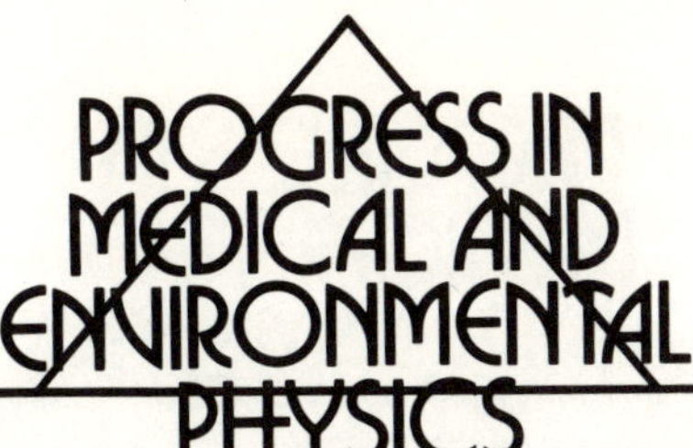

PROGRESS IN MEDICAL AND ENVIRONMENTAL PHYSICS

VOLUME 2

IMAGING WITH NON-IONIZING RADIATIONS

General Editor

Daphne F. Jackson, D.Sc., F.Inst. P., MIEEE

Honorary Adviser

W.V. Mayneord, F.R.S., C.B.E., D.Sc., D.Univ. (Surrey)

This series presents critical reviews of current developments in the applications of physical techniques to medicine, the life sciences and the physical and environmental sciences. Each volume concentrates on a particular area of physics. The series provides authoritative guidance for academic and industrial scientists and engineers on the potential of new techniques and the fundamental science on which they are based.

Other titles in the series:

Volume 1 *Imaging with Ionizing Radiations*
(Kouris, Spyrou and Jackson)

LEICESTER POLYTECHNIC
LIBRARY
Acc. No.
Date 25.9.84
Loc./Form
Class. 616.0754
IMA

Edited by

Daphne F. Jackson, D.Sc., F.Inst. P., MIEEE

Professor of Physics

University of Surrey

Surrey University Press

Published by Surrey University Press
A member of the Blackie Group
Bishopbriggs, Glasgow G64 2NZ and
Furnival House, 14–18 High Holborn, London WCIV 6BX

© 1983 Blackie & Son Ltd.
First published 1983

All rights reserved. No part of this publication may be reproduced, stored in a retrieval system, or transmitted, in any form or by any means, electronic, mechanical, recording or otherwise, without prior permission of the Publisher

Distributed in the United States by
Sheridan House Inc., New York
ISBN 0-911378-46-4

British Library Cataloguing in Publication Data

Imaging with non-ionizing radiations.—(Progress in medical and environmental physics; v. 2)
1. Imaging systems in medicine 2. Diagnosis
I. Jackson, Daphne F. II. Series
616.07′54 R857.06

ISBN 0–903384–37–X

Filmset by Thomson Press (I) Limited, New Delhi,
Printed in Great Britain by Bell & Bain Ltd, Glasgow

CONTRIBUTORS

P.S. Allen, Ph.D., F.Inst. P.
Department of Applied Sciences in Medicine,
University of Alberta, Edmonton, Canada.

Brian H. Brown, B.Sc., Ph.D., F.Inst. P., MIEE C. Eng.
Department of Medical Physics and Clinical Engineering,
Royal Hallamshire Hospital, Sheffield S10 2JF.

R.C. Chivers,
M.A., Ph.D., F.I.O.A., A.F.I.M.A., F.Inst. P., M.Inst. N.D.T., F.R.S.A.
Department of Physics,
University of Surrey, Guildford, Surrey GU2 5XH.

Daphne F. Jackson, D.Sc., F.Inst. P., MIEEE,
Department of Physics,
University of Surrey, Guildford, Surrey GU2 5XH.

Colin H. Jones, B.Sc., Ph.D., F.Inst. P.
Department of Physics, The Royal Marsden Hospital and
Institute of Cancer Research,
Fulham Road, London SW3 6JJ.

Kypros Kouris, Ph.D., M.Inst. P.
Department of Radiology and Nuclear Medicine, Faculty of Medicine,
Kuwait University, Kuwait.

D. Watmough, B. Sc., Ph.D., F.I.O.A.
Department of Biomedical Physics and Bioengineering,
University of Aberdeen, Foresterhill, Aberdeen AB9 2ZD.

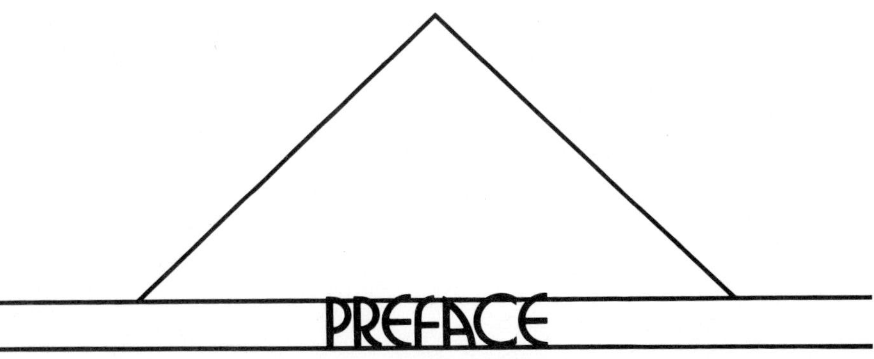

PREFACE

This book is the second in a series of monographs published under the general title *Progress in Medical and Environmental Physics.* These monographs are intended to give an account of new developments which are making an impact on the application of physical techniques in medicine, the life sciences and environmental sciences, and, in some cases, also in materials sciences, certain aspects of engineering, and archaeometry. They are aimed primarily at graduate scientists and engineers who are entering the field, but may also be useful to senior undergraduates taking special options or carrying out projects in medical or environmental physics. Experienced workers in these fields should find the monographs useful as critical reviews of current developments. It is not intended to describe in detail applications in clinical practice, but certain volumes may nevertheless be of value to clinicians who wish to evaluate the potential of new techniques. For example, the volumes on imaging should be of interest to diagnostic radiologists, and to nuclear medicine physicians. Similarly, routine industrial applications will not be discussed but new techniques relevant to analysis and non-destructive testing will be covered.

This book is concerned with imaging using non-ionizing radiations. It complements the first volume which covered imaging with ionizing radiations, so that descriptions of image reconstruction techniques and the relevant mathematics have not been repeated. Chapter 1 reviews new imaging techniques using both ionizing and non-ionizing radiations, and provides a link between Volumes 1 and 2. Chapter 2 deals with the well-established technique of imaging with ultrasound, but addresses an issue of current interest, namely quantitative analysis of the texture of the images produced. In Chapter 3, the electrical properties of tissue are described, as this is essential information for the development of imaging by impedance and NMR measurements. Chapter 4 concentrates on the techniques of NMR imaging (its applications are included in Chapter 1) while Chapter 5 covers a wide range of topics in thermal imaging, with emphasis on the infrared region. Some interesting non-medical applications are included. Finally, Chapter 6 gives an account of an optical technique of imaging, applied particularly to imaging of the breast.

I am indebted to the authors of the separate articles in this volume for their co-operation and dedication. Appropriate acknowledgements appear at the end of each article, but my thanks are due to Professor W. V. Mayneord, Honorary Adviser for the series, for this sympathetic support and encouragement, and to John Grant for his assistance with the preparation of this volume.

D. F. J.

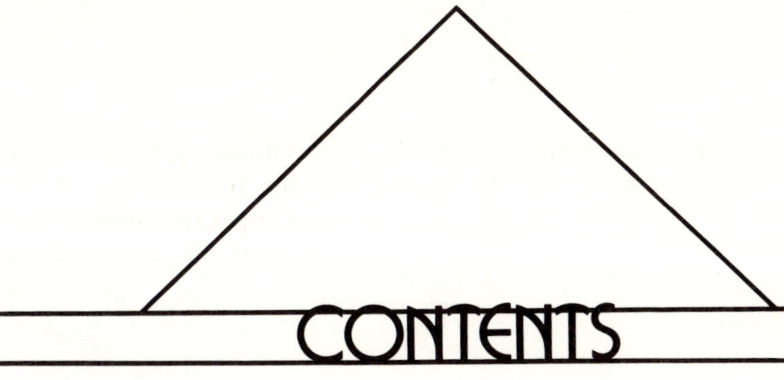

CONTENTS

Daphne F. Jackson and Kypros Kouris

1.1 Introduction

1.1.1 *Objective—the choice of appropriate techniques*

Imaging is concerned with the assessment of the morphology and/or function of objects that are usually not accessible for direct visual inspection. Objects of interest are found both in medicine and industry giving rise to a wide class of imaging applications such as biomedical imaging, micrography of molecules and crystals, space imaging and radioastronomy, satellite mapping of earth resources and reconnaissance photography, radar mapping, formation of tracks of particles and ions, and non-destructive testing of materials. In general, the objective of an imaging investigation is to derive information on the spatial and time variations of a property, or combination of properties, of the object under study. The information is in the form of an image, or a series of images, possibly together with a set of values of parameters reflecting the state of the object. On the basis of this information, the observer proceeds to take decisions concerning, for example, the state of health of a patient and the patient's management, or the presence of a crack or cavity in an aircraft wing (Volume 1, chapters 1, 2).

In the following sections of this chapter, some imaging methods are compared according to their ability to provide information on the morphology, function, or physical and chemical characteristics of the object. In medical diagnosis, imaging methods have been used primarily to study morphology and function. This need not imply that analytical information is not required, but that it is frequently obtained by independent measurements, e.g. from blood and urine samples or by biopsy. Connections between imaging and analysis are developing, however, and obvious examples are CT-guided biopsy, tissue characterization with ultrasound and X-ray CT, and biochemical studies with NMR. For studies of tissue samples *in vitro* and of various biological materials, analysis is usually of primary interest but is often carried out by scanning techniques which map the distribution of the constituents and hence provide an image on a microscopic scale. For these samples, techniques can be used which are effective only for surfaces or thin sections. For all biological and botanical materials, radiation damage frequently imposes constraints and for studies *in vivo* there are further restrictions arising from trauma, and both regular and involuntary movements. In non-destructive testing of inanimate objects, dose consideration may not form a constraint. Involuntary movement does not normally occur but regular motion up to quite high speeds may be involved, for example of turbine blades, objects on a conveyor belt, or in fluid flow. The irregular shape of such objects also affects the choice of reconstruction algorithm (1). Precise location of a defect may not be important but detection of defects of small dimension is a primary requirement.

An imaging method is specified by the type of radiation used and by the imaging system employed. There is a wide range of methods and techniques, and yet for a given application there is not necessarily an optimum choice; even if there is such a choice the task of selecting it can be difficult. The comparisons given here will emphasize resolution, sensitivity, specificity, and also ability to provide information not otherwise available. The importance of the last two properties can be appreciated from the following medical case history (2). Radionuclide liver-spleen scintigrams of a female patient, obtained 5 months after resection of an adenocarcinoma of the sigmoid colon, were interpreted as normal. However, her carcinoembryonic antigen (CEA) level was elevated to 82 ng/ml and, a month later, increased to 90 ng/ml. During further surgery, an inflamed region of the sigmoid colon was resected but all abdominal organs were found normal. Her CEA levels 12 and 18 months after initial surgery were 235 ng/ml and 542 ng/ml, respectively, and a month later, although a liver-spleen scan was again found to be normal, her CEA level was greater than 1250 ng/ml. An extensive diagnostic evaluation was then undertaken by radionuclide bone scan, liver-spleen scan, liver biopsy, ultrasound of the pelvis, barium enema and X-ray CT; all were normal but her CEA level had increased to 2850 ng/ml. A tumour localization study was then undertaken—two days after the intravenous administration of 37 MBq of ^{131}I-labelled antibodies to CEA, the patient received 37 MBq of ^{99m}Tc-labelled albumin. Abnormal concentration of radioactivity was observed in the liver and above the urinary bladder. Because of these findings, the patient was given 19 MBq of ^{99m}Tc sulphur colloid which confirmed the liver abnormality. A large tumour was found at surgery, and a liver biopsy confirmed metastatic adenocarcinoma.

Several conclusions may be drawn from this medical case. Although established imaging techniques often provide complementary information, their failure to detect the hepatic lesion is a demonstration of their non-specificity; conversely, the specificity of the CEA antibody study towards neoplasms containing CEA is the fundamental reason for its success. The general approach of identifying specific antigens and their antibodies, developing labelled compounds and applying appropriate imaging procedures, is termed *in-vivo* radioimmunodetection and is believed to have significant potential (3). The continuous rise of the CEA level was a consistent indication of disease although its general utility as such is, at present, not sufficiently well established. However, it may be that in current research with the emphasis laid on imaging methods, not enough effort is being directed towards deriving single-valued disease indicators.

1.1.2 *Summary of imaging methods*

Formally, imaging is an indirect sensing problem (4, 5): the mathematical function $f(\mathbf{r}, t)$ is sought but can only be inferred from measurements $p(\mathbf{r}, t)$, such that

$$p(\mathbf{r}, t) = \int d\mathbf{r}' \, f(\mathbf{r}', t) K(\mathbf{r}, \mathbf{r}', t) \tag{1.1}$$

where the kernel $K(\mathbf{r}, \mathbf{r}', t)$ expresses the measurement process. In imaging, the kernel reflects the type of radiation used, its interactions with the object and the detectors, the measurement geometry, and the geometry of the object. For a given z and t, and under the assumption that during the measurement interval Δt the property of interest does not change significantly, the problem reduces to inferring an estimate $\hat{f}(x, y)$, the image.

A further generalization is to describe the imaging process in terms of an input-output system characterized by an operator or mapping T (6, 7), where

$$\hat{f} = Tf. \tag{1.2}$$

In tomography, the image $\hat{f}(x, y)$ must be a close representation of the tomographic plane $f(x, y)$ and must be free of contributions from adjacent planes. The overall impulse response function $h(\mathbf{r}', x, y)$ expresses mathematically how the imaging system transforms a δ-function input $\delta(\mathbf{r}')$ through the stages of data collection, reconstruction and processing. Thus

$$\begin{aligned} h(\mathbf{r}', x, y) &= T\delta(\mathbf{r}') \\ \hat{f}(x, y) &= \int d\mathbf{r}' \, f(\mathbf{r}') h(\mathbf{r}', x, y). \end{aligned} \tag{1.3}$$

The modulus of the Fourier transform of $h(\mathbf{r}, x, y)$ is the modulation transfer function (MTF).

Within the above framework, image reconstruction from projections is concerned with the reconstruction of images from measurements which provide estimates of the projections of the object at different orientations

(Vol. 1, chapter 3; 8). If the object, assumed to be of finite extent, is a plane, its projections are one-dimensional; if the object is 3-D then, depending on the measurement process, 1-D or 2-D projections can be obtained. In two dimensions, and using polar co-ordinates, the function $f(r, \phi)$ can be uniquely reconstructed from its projections $p(l, \theta)$ where

$$p(l, \theta) = \int_{L(l,\theta)} ds\, f(r, \phi) \tag{1.4}$$

provided that $f(r, \phi)$ is continuous, bounded and of compact support, and the line integrals $p(l, \theta)$ are known for all lines $L(l, \theta)$ traversing the object. Then, the inversion formula is

$$f(r, \phi) = -\frac{1}{2\pi^2} \int_0^\pi d\theta \int dl \frac{\partial}{\partial l} p(l, \theta) \frac{1}{(l - r\cos(\phi - \theta))} \tag{1.5}$$

which, on integrating by parts and taking account of the singularity, becomes

$$f(r, \phi) = \frac{1}{2\pi^2} \int_0^\pi d\theta \lim_{\varepsilon \to 0} \int dl\, p(l, \theta) q_\varepsilon(l - r\cos(\phi - \theta)) \tag{1.6}$$

where

$$q_\varepsilon(s) = \begin{cases} 1/\varepsilon^2 \text{ if } |s| < \varepsilon \\ -1/s^2 \text{ otherwise.} \end{cases} \tag{1.7}$$

In three dimensions, and provided that the function $f(\mathbf{r})$ obeys the above conditions, $f(\mathbf{r})$ can be uniquely reconstructed from knowledge of either the planar integrals or the line integrals over all planes or lines, respectively, traversing the object. If the projection data are planar integrals then the inversion formula is (9)

$$f(\mathbf{r}) = -\frac{1}{2\pi} \nabla^2 g(\mathbf{r}) \tag{1.8}$$

where $g(\mathbf{r})$ is the mean value of the integral over all the planes which traverse an arbitrarily small sphere centred at $\mathbf{r}$. If the projection data are line integrals then one method of reconstruction is a direct generalization of ART (8).

The transition from mathematical theory demanding infinite sampling to practical implementations with finite sampling must comply with the sampling theorem. In two dimensions, the spatial and angular sampling intervals must be given by (Vol. 1, chapter 3; 10)

$$\Delta l \lesssim d/2; \quad \Delta\theta \lesssim 2d/D \tag{1.9}$$

where d is the pixel dimension, D the diameter of the reconstruction region, and $\Delta\theta$ is in radians.

Imaging methods employ radiation in one of three modes: transmission,

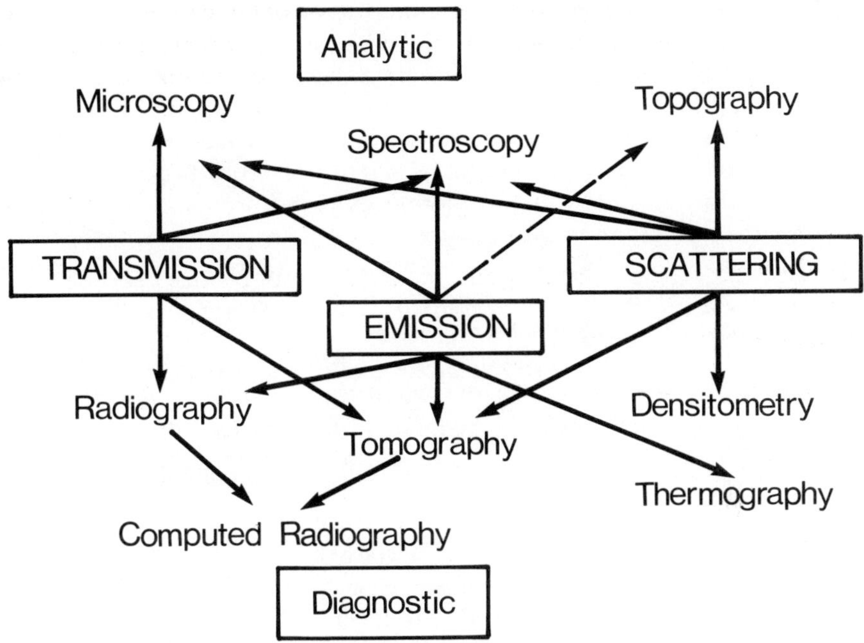

Figure 1.1 Analytic and diagnostic techniques based on transmission, emission and scattering.

emission, and scattering, as indicated in figure 1.1 (11). Methods in the upper half of thc diagram are primarily analytic while those in the lower half are primarily diagnostic.

Transmission methods depend on the differential absorption of the radiation used. For image interpretation and for extraction of quantitative information it is necessary to understand the pertinent energy loss mechanisms and the physical parameters on which they depend.

The energy loss of photons is exponential with depth and the attenuation coefficient $\mu(r, \phi, E)$ at energy E is the sum of contributions from the photoelectric effect, elastic scattering, inelastic scattering, and pair production (12, 13). For an incident monochromatic photon beam of intensity $I_0(E)$, measurement of the transmitted intensity $I(E)$ along the line $L(l, \theta)$ gives the projection estimate

$$p(l, \theta, E) = -\log_e (I(E)/I_0(E)) = \int_{L(l,\theta)} ds\, \mu(r, \phi, E) \tag{1.10}$$

and image reconstruction would yield the distribution of linear attenuation coefficients in the inhomogeneous object of interest. The use of a γ-ray source emitting photons at a number of discrete energies, together with an energy sensitive detector, can give a set of projections and hence a set of images, one at each energy. However, if a polychromatic X-ray beam is used, as in X-ray CT scanners, image interpretation must take account of the beam hardening problem (14).

Because the energy loss for neutrons is also exponential with depth, the mathematical description of image reconstruction from projections using neutrons is fundamentally as for photons. A monochromatic beam can be accomplished by inserting into a reactor beam a material which exhibits a neutron window, i.e. absorbs all neutrons except those of energy within a narrow range (15). In this case, the equation for image reconstruction is of the same form as eq. (1.10). The neutron attenuation coefficient is strongly dependent on isotopic composition and neutron energy, and, in contrast to photons, there is no systematic variation with atomic number. This implies that, for each object to be examined, there may be an optimum neutron energy, or neutron spectrum, and a matching detector, or set of detectors (16).

The main energy loss mechanism for charged particles is ionization. The rate of energy loss (stopping power) is given by the Bethe-Bloch formula (Volume 1, chapter 5). For a mixture of elements, the Bragg additivity rule is used, corresponding to the mixture rule for photon attenuation coefficients (see section 1.5.1). The dominant physical property in the energy loss mechanism is the electron density and the charged particles have a well-defined range depositing most of their energy at the end of the range. Residual range or residual energy measurements can be converted to an effective thickness $p(l, \theta)$ of material penetrated using the stopping power (17), and this is proportional to the line integral of the density $d(r, \phi)$ along the trajectory $L(l, \theta)$, i.e.

$$p(l, \theta) \propto \int_{L(l,\theta)} ds\, d(r, \phi), \tag{1.11}$$

which is of the same form as eq. (1.4). In contrast to X-ray CT, there is no beam hardening, but multiple scattering leads to divergence of the beam so that, strictly, $L(l, \theta)$ represents a nonlinear path.

Nonlinear, curved paths arise in the propagation through matter of ultrasound, electric current and microwaves (18, 19), rendering image reconstruction a nonlinear problem (20). The linear discretized problem can be represented by a linear set of equations

$$p = Kf \tag{1.12}$$

where f and p are column vectors representing the unknown function and the projection data respectively, and K is a known matrix dependent on the geometry but not on the object. The nonlinear discretized problem is correspondingly represented by

$$p = K(f)f \tag{1.13}$$

where the matrix $K(f)$ is dependent on both the geometry and on f. Thus the paths along which the radiation is propagated are curved, object dependent, and not always confined to the tomographic slice. Although a generalization of ART can give a solution, a good initial guess of f is required (20).

In a homogeneous medium the propagation of ultrasound is completely characterized by the attenuation coefficient and the refractive index (or speed), but in an inhomogeneous medium reflection and refraction also occur due to differences in acoustic impedance (21). There are two general approaches to ultrasound CT: one is based on attenuation measurement and gives the distribution of acoustic attenuation coefficients, and the other is based on time-of-flight measurements and gives the distribution of acoustic refractive index. In both approaches the refractive index $n(r, \phi)$ is needed if curved paths are not to be ignored. The path $L(l, \theta, n)$ traversed by a pulse must, by Fermat's principle, be associated with the minimum propagation time, and is therefore a function of the refractive index. The time-of-flight measurement $p(l, \theta)$, is then given by (22)

$$p(l, \theta) \propto \int_{L(l,\theta,n)} ds\, n(r, \phi). \tag{1.14}$$

If curved paths are ignored then eq. (1.14) reduces to

$$p(l, \theta) \propto \int_{L(l,\theta)} ds\, n(r, \phi) \tag{1.15}$$

where $L(l, \theta)$ is the straight line defined by the transmitting and receiving transducers. Eq. (1.15) is of the same form as eq. (1.4). A fundamental limitation of ultrasound CT, irrespective of the imaging approach, is that regions exist which can not be intersected by any minimum propagation time paths (23); since such a region may surround other regions, they too can not be faithfully reconstructed. In contrast to attenuation ultrasound CT, the time-of-flight approach is free from errors due to reflection and refraction processes.

Curved paths and non-uniqueness problems are important limitations in inferring the resistivity distribution $p(\mathbf{r})$ from external resistance measurements (20, 24). In particular it is difficult to image a high resistivity region surrounded by a region of high conductivity (25). The streamlines of the electric current follow the gradient of the potential field, and the reconstruction problem can not generally be reduced to two dimensions, although the use of guarded electrodes is a possible approach (26). In general, the column vector f, defined in eq. (1.13), would contain the discretized 3-D representation of the resistivity $\rho(\mathbf{r})$; each element of the column vector p would be given by the ratio of the applied voltage to the measured current, the path traversed by the current being dependent on the unknown $\rho(\mathbf{r})$ and the matrix $K(f)$ would depend on $\rho(\mathbf{r})$, so that if $\rho(\mathbf{r})$ were known, column p would be calculable.

Microwave transmission CT is thought to exhibit high sensitivity for applications involving soft tissues, such as lungs, where ultrasound is highly attenuated and dispersed (19). The range of radio and microwave frequencies investigated is 10 MHz–10 GHz but better resolution is expected at the higher frequencies, particularly if the phase rather than the amplitude of the transmitted signal is used. At 10.5 GHz, 2 cm resolution has been achieved by assuming linear paths and exponential attenuation, and neglecting reflection and diffraction processes (27); under these assumptions, image reconstruction is as described by eq. (1.4). At lower frequencies, degradation of image quality due to nonlinear propagation is expected to be significant. However, satisfactory reconstructions of attenuation and refractive index have been obtained using data collected by cross-borehole electromagnetic probing at 50 MHz, assuming linear paths (28).

Emission methods rely on the external detection of internally emitted radiation. In the passage of the radiation through the medium, from the point of emission to the detector, physical interactions may occur, including absorption and scattering. For quantitative emission imaging, detection of the radiation along multiple trajectories is necessary.

Radionuclide emission imaging is concerned with the determination of the activity distribution $a(\mathbf{r}, t)$ which describes the uptake, localization and turnover of the administered labelled compound in the region under study (Vol. 1, chapters 6 and 7); the radionuclide can be either a γ-ray or a positron-emitter. In medicine, emission imaging provides primarily physiological information (29); dynamic studies can follow the metabolism of the radiopharmaceutical and metabolic parameters can be derived on application of compartmental analysis (30).

With γ-ray emitters, it is necessary to use coded apertures (31, 32) or collimators in order to define photon trajectories and thereby relate a point of detection with a point, or region, of emission. There are two general geometries in γ-ray emission tomography depending on whether the photon trajectories traverse several tomographic planes, or are constrained in a single slice. In the former case, the data contain 3-D information about the activity distribution, but in a coded form which depends on the coded aperture or collimator used. The angular sampling is necessarily limited and hence tomographic images are dependent on object orientation; out of plane contributions lead to artefacts (33, 34). In the latter case, the geometry is that of 2-D image reconstruction from projections. The intensity of photons detected along the line $L(l, \theta)$ is given by

$$p(l, \theta) = \int_{L(l,\theta)} ds\, a(r, \phi) \exp\left[- \int_{L(l,\theta,r,\phi)} ds\, \mu(r', \phi') \right] \tag{1.16}$$

where $L(l, \theta, r, \phi)$ denotes the line segment of $L(l, \theta)$ from the point of emission (r, ϕ) to the detector. With $\mu(r', \phi')$ denoting the linear attenuation coefficient, the inner term represents the exponential attenuation of the emitted photons.

With positron emitters, the coincidence detection of the annihilation γ-rays signifies that a labelled molecule existed somewhere along the line joining the two opposing detectors; strictly, this interpretation is valid provided that the range of the positron and the departure from collinearity of the two photons can be neglected (Vol. 1, chapter 7). The coincidence measurement along the line $L(l, \theta)$ is given by

$$p(l, \theta) = \int_{L(l,\theta)} ds\, a(r, \phi) \exp\left[- \int_{L(l,\theta)} ds\, \mu(r, \phi) \right] \tag{1.17}$$

where it is evident that, in contrast to eq. (1.16), the attenuation integral is not dependent on the point of emission. For each line $L(l, \theta)$, the attenuation term $c(l, \theta)$ may be regarded as a constant, and if it were known eq. (1.17) would reduce to

$$p(l, \theta)/c(l, \theta) = \int_{L(l,\theta)} ds\, a(r, \phi) \tag{1.18}$$

which is of the same form as eq. (1.4). With a coincidence time window less than the photon transit time in the object of interest, time-of-flight positron tomography can lead to significant improvements in the image signal/noise ratio (35, 36); alternatively, for a given signal/noise ratio the dose can be reduced or faster dynamic studies can be performed.

Methods for attenuation correction in emission tomography with γ-ray or positron emitters involve either an independent set of transmission measurements, or calculations based on an assumed (commonly uniform) attenuation distribution (37, 38).

Human tissue in thermal equilibrium emits radiation as a result of internal thermal motion. The power $I(\nu)$ emitted through unit area into a unit solid angle in a frequency interval ν, $\nu + d\nu$ is given by Planck's law (see chapter 5)

$$I(\nu) = \varepsilon(2h\nu^3/c^2)[\exp(h\nu/kT) - 1]^{-1} \tag{1.19}$$

where T is the absolute temperature. For radiation by the human skin into air at wavelength $\lambda = 10\ \mu$m, the emissivity ε is close to its black body value of unity and the emitted intensity is maximum (39, 40). The external detection of the emitted infrared radiation, in infrared thermography, gives information about the surface temperature since, at this wavelength, the penetration depth in tissue is ~ 1 mm. Spatial resolution is ~ 1 mm and temperature resolution ~ 0.2 K.

At microwave frequencies, Planck's law reduces to the Rayleigh-Jeans law

$$I(\nu) = \varepsilon \cdot (2kT\nu^2/c^2) \tag{1.20}$$

so that, in contrast to infrared thermography, the detected intensity in microwave thermography is directly proportional to the absolute temperature. At $\lambda = 0.1$ m, $\nu = 3$ GHz, the emissivity ε is 0.5 and, although the intensity is $\sim 10^{-8}$ of that emitted at infrared frequencies it can be detected with a temperature resolution of ~ 0.1 K (40, 41). The absorption of microwaves depends on water content: it is least for low water content, e.g. fat (penetration ~ 50 mm), and highest for high water content, e.g. muscle (penetration ~ 10 mm). Spatial resolution is ~ 10 mm.

Quantitative thermography requires careful control of instrumentation, ambient conditions and patient preparation. The thermographic index expresses the mean temperature difference of the region of interest from a constant, the constant being the mean regional temperature of control groups measured under identical conditions.

Nuclear magnetic resonance (NMR) is a process of resonant absorption and re-emission of radiofrequency (RF) energy (chapter 4). The interactions between the externally applied static magnetic field and the nuclei excited by the RF field, provide information about the physical and chemical environment in which the nuclei exist, and their modes of interaction with their surroundings (42–45). The magnetization vector is subject to two relaxation processes (chapter 4)—the spin-lattice relaxation, characterised by a time constant T_1, and the spin-spin relaxation, characterised by a time constant T_2.

NMR imaging provides spatial information by using magnetic field gradients in such a way that during the imaging time interval each point in the sample experiences a unique magnetic field history. A linear magnetic field gradient can be written as $\mathbf{G}_r = G\hat{\phi}$ and the total field becomes $\mathbf{B}(\mathbf{r}) = \hat{k}(B_0 + G\hat{\phi}\cdot\mathbf{r})$ where $\hat{\phi} = \hat{\imath}\cos\phi + \hat{\jmath}\sin\phi$. The NMR signal is then given by (45)

$$V(t,\phi) = C\int d\tau \exp(i\gamma G\hat{\phi}\cdot\mathbf{r}t) M_+(\mathbf{r},0) \tag{1.21}$$

$$= C\int dr \int d\theta |r| \exp(i\gamma G\hat{\phi}\cdot\mathbf{r}t) \int dz\, M_+(\mathbf{r},0) \tag{1.22}$$

where $d\tau = dr\,d\theta\,dz|r|$ in cylindrical polar coordinates, γ is the gyromagnetic ratio and

$$\int dz\, M_+(\mathbf{r}, 0) = f(r, \theta) \tag{1.23}$$

is the projection of the transverse magnetization at time $t = 0$ on to the plane $z = 0$. Thus the NMR signal $V(t, \phi)$ is the 2-D Fourier transform of this projection. Signals are measured with gradient directions in the range $0 \leq \phi < \pi$ and the inverse Fourier transform is taken, to give

$$f(r, 0) = \int dt \int d\phi |t| \exp(-i\gamma G\hat{\phi}\cdot\mathbf{r}t)V(t, \phi) \tag{1.24}$$

$$= Cf(r, \theta) \tag{1.25}$$

This is the NMR image, which is directly connected to the NMR signal by a 2-D Fourier transform. The 1-D projection of the transverse magnetization is given by

$$p(r, \phi) = \int dt \exp(-i\gamma Grt)V(t, \phi) \tag{1.26}$$

$$= C \int d\tau\, M_+(\mathbf{r}, 0) \tag{1.27}$$

since $\hat{\phi}\cdot\mathbf{r} = r$. This projection is completely analogous to those discussed earlier so that the same reconstruction techniques can be applied (45, chapter 4).

The different methods of excitation lead to different forms for the transverse magnetization. For the 90° free precession method, the transverse magnetization is given by (45, 46)

$$M_+^{fp} = -iM_0(1 - \varepsilon_1)/(1 - \varepsilon_1\varepsilon_2) \tag{1.28}$$

where M_0 is the equilibrium magnetization, $\varepsilon_1 = \exp(-T_r/T_1)$, $\varepsilon_2 = \exp(-T_r/T_2)$, and T_r is the repetition interval. If T_r is very long compared with T_1 and T_2, the image is a map of the equilibrium magnetization and is usually called a proton density map. For the inversion recovery method (chapter 4), the transverse magnetization immediately after the 90° pulse is (45, 46)

$$M_+^{ir} = iM_0(1 - 2e_1 + \varepsilon_1')/(1 - \varepsilon_1'\varepsilon_2') \tag{1.29}$$

where the repetition interval is T_r', $e_1 = \exp(-\tau/T_1)$, and τ is the recovery time. If the repetition times are long compared with the relaxation times, the ratio M_+^{ir}/M_+^{fp} is given by

$$M_+^{ir}/M_+^{fp} \sim -(1 - 2e_1 + \varepsilon_1')/(1 - \varepsilon_1) \tag{1.30}$$

which is dependent only on T_1. This ratio is used to derive maps of T_1. The transverse magnetization given by the spin-echo method (chapter 4) is given by (45, 46).

$$M_+^{se} = -ie_2^2(1 - 2E_1 + \varepsilon_1')/(1 - \varepsilon_1'\varepsilon_2') \qquad (1.31)$$

where $E_1 = \exp(-T_r/T_1 + \tau/T_1)$ and $e_2 = \exp(-\tau/T_2)$. If the repetition rates are nearly equal and pulse separation is small by comparison with the repetition time, the ratio M_+^{se}/M_+^{fp} is proportional to e_2^2, which depends only on the transverse relaxation time T_2. This ratio is used to derive maps of T_2. Other methods can be used to derive M_0, T_1 and T_2 (44, 47, 48), and are based on the same analysis of the magnetization. Both relaxation times may be changed by paramagnetic materials which can act as contrast agents (49).

NMR is not a sensitive technique and it is not yet clear whether response of tissue is specific (50); it seems that both T_1 and T_2 images are needed for diagnosis (50, 51). In general, the intensities on an NMR image do not show an unambiguous relationship to tissue properties, and the use of simple models of the response can be misleading especially if these are derived from the response of excised tissue (45).

Through measurement of chemical shifts of the nuclear resonance line, and the splitting pattern of the line, chemical and biochemical information is derived. When this information is obtained from a small localized volume, the process is known as topical NMR or TMR. The information which can be obtained from TMR is discussed further in section 1.5.2 of this article.

Table 1.1 Comparison of NMR imaging and TMR (204).

NMR imaging	*TMR*
Inhomogeneously broadened spectra (usually ^1H although some work on ^{31}P).	High resolution spectra (usually ^{31}P)
Low field ($< 0.4T$) Low homogeneity (1 in 10^5)	High field ($2T$) High homogeneity (1 in 10^7)
Switched magnetic field gradients and/ or radiofrequency field gradients.	Static field gradients
The final image contains NMR signals acquired from the whole sample	NMR signals acquired only from a selected volume of the sample.
Spatial information in the form of two (or three) dimensional maps of nuclear density, T_1 or T_2 relaxation times,* is produced.	Biochemical information is obtained from the resonance frequencies and the chemical shifts.

*or a combination of these properties.

A comparison of the features of NMR imaging and TMR is given in table 1.1 (204). Although, currently, NMR imaging work concentrates on hydrogen, research is being directed towards other magnetic nuclei such as ^{31}P. The use of NMR contrast agents such as Mn^{2+}, ^{17}O and ^{19}F is also under investigation. The use of stable isotope tracers to investigate metabolic function in a manner similar to radionuclide studies should be feasible.

1.2 Morphological information

1.2.1 *Anatomy of internal structures*

The morphological information of primary importance in medical diagnosis is the anatomy of normal internal structures in the body, especially in the vicinity of any suspected abnormality. For radiotherapy treatment planning, it is necessary to delineate regions of particularly low or high absorption of radiation, e.g. air cavities or bone. Recognition of well-defined surfaces may be an important aid to diagnosis, e.g. observation of encapsulation or some other borderline, and also shift of the mid-line of the brain. Hence, it is necessary that the physical response of the body to the probe should serve to differentiate between the various regions and structures or should change at interfaces. The information required for the general investigation of non-medical objects is quite similar but for non-destructive testing the ability also to detect cracks, voids and inclusions down to rather small dimensions is needed. In this field of investigation the spread of object sizes and densities of interest can be very great.

Ultrasound measures the anatomical position of surfaces and tissue structure characterization arises through differences in acoustic impedance. Structures of a few mm in dimension can be resolved in soft tissue in the interior of the body, while resolution of ~ 0.1 mm can be achieved in superficial regions in inanimate objects and hence crack detection is possible (52). However, ultrasound is strongly reflected at boundaries between soft tissue and gas or bone and ultrasound which does penetrate into gas or bone suffers high attenuation (53). Thus an important medical application is the study of soft tissue, especially in the abdominal cavity.

Ultrasound also has high potential for observation of the motion of surfaces, such as heart valves and chambers (54, 55). Real-time scanning gives particularly good information on spatial relationships within the heart, whereas M-mode echocardiography is better suited to detailed investigation of valve or wall motion (55, 56). Real-time scanning also has potential for visualization of intracranial structures, including cerebral structures (56).

X-ray CT has proved particularly successful in providing anatomical information, especially for the brain where it has largely replaced invasive techniques. Because of the high contrast which arises between bone and soft tissue, X-ray CT reveals the anatomy of the spine in great detail (57). This method is sensitive, gives good resolution, but is generally not specific. This lack of specificity leads to difficulties in characterization (see section 1.2.3) unless additional measurements are taken at different energies (see section 1.4.3) which in turn increases the dose.

The application of X-ray and γ-ray CT to non-medical objects has recently been reviewed. Not surprisingly, use of radionuclide sources emitting X-rays in the same energy range as a CT X-ray tube is most successful for wood, rock samples saturated with oil or water, pottery and botanical specimens (58, 59). Low density materials can be imaged very well at ~ 18 keV (59) while much higher photon energies are required for thicker, more dense, objects (1, 60). Special techniques can be adopted to ensure uniform statistical errors and this enables confidence limits to be set for detection of cracks, voids and inclusions, which may be detected even though their dimensions are well below the pixel size (14). Stress corrosion cracks < 0.3 mm wide have been observed, and holes, voids and inclusions down to 0.25 mm have been detected (1, 60).

Charged particle beams essentially image density variations. High resolution can be achieved at significantly lower doses than X-ray CT, but these methods also lack specificity. Improved visualization of soft tissue structures, such as tendons, may prove to be of clinical significance (61) and there may be valuable non-medical applications (62).

In NMR imaging there is a choice of parameters which may be displayed. These include the concentration of nuclei of a chosen species (usually hydrogen), the T_1 relaxation time and the T_2 relaxation time. NMR proton distributions can be obtained with high spatial resolution, of ~ 1 mm for the whole body.

NMR does not suffer from bone artefacts and therefore has an advantage over X-ray CT in the posterior fossa of the brain, which is a common site for multiple sclerosis lesions (63). NMR proton distributions reveal the anatomical distribution of bone particularly clearly but it is thought unlikely that NMR imaging will displace X-ray CT for investigation of the spine and its contents (64) and detailed bone changes associated with disease are not readily observed (65). Patient movement is not as great a problem with NMR as with X-ray CT; the interface between grey and white matter is an additional source of partial volume effects but this interface also provides useful anatomical landmarks (63).

Non-medical applications of NMR imaging include any situation where distribution or amount of water content is important, e.g. state of preservation of food grains, distribution of liquid in porous materials and interfaces between water and other media.

1.2.2 *Visualization of the vascular system*

An important requirement in morphological diagnosis is the accurate delineation of the vascular system and the indication of changes in vascular anatomy and pathology. A knowledge of vessel architecture is needed for surgical planning.

Comprehensive topographic surveys of vascular areas can be obtained from conventional angiography, which involves the injection of contrast material via arterial catheters. More recently, digital subtraction angiography has provided a non-invasive method using intravenous injection of contrast material. With subtraction of the digitized background image due to tissue alone from the enhanced image of tissue plus contrast material, a high contrast image is achieved. This method utilizes a conventional X-ray tube but requires a high quality image intensifying system together with a video camera of high dynamic range (66, 67). This method is very effective and can be applied to peripheral arteries, arterial and venous circulation of the brain and pulmonary vasculature. Difficulties arise when there is superposition of other signals which distort the information required, e.g. in imaging the coronary vascular tree (54).

Dynamic computed tomography can be used to follow the distribution of contrast material in the larger arterial and venous vessels (68). The CT scanner must operate with short scan times and high scanning frequency. Measurements can be made at different slice levels, for example to represent aortocoronary bypass vessels (69). Vascular structures can be represented in different perfusion states, thereby relating morphological to functional information.

Direct visualization of blood vessels is possible using Doppler ultrasound systems, in particular using Doppler imaging systems and duplex scanners (56, 70). The carotid and opthalmic arteries may be studied in this way (71). For stenoses greater than 50% there is excellent agreement between Doppler methods and arteriography (72), but below 50% pulsed Doppler systems are preferred due to their ability to provide 3-dimensional visualization. The same methods can be applied to the peripheral vascular system (71). NMR imaging also offers considerable potential for detection of vascular pathologies, and for monitoring and characterization of deposits in diseased vessel walls (54). The application of NMR and ultrasound and in blood flow perfusion studies is discussed in section 1.3.2.

1.2.3 *Visualization and characterization of abnormalities*

A few important abnormalities are discussed here in order to compare the potential of various methods.

Neoplastic abnormalities normally appear as zones of decreased density with significant associated mass, and the zone appears to surround a focal mass (57). (There are many exceptions to this simplified description.) Such abnormalities are recognised in X-ray CT on the basis of the initial image and changes in the CT values with time, but CT does not distinguish very effectively between solid and cystic tumours. Ultrasound detects tumours due to differences in acoustic impedance and internal echo patterns and gives a clear distinction between solid and liquid regions (53, 73). The texture of ultrasound patterns is partly due to tissue structure (histology) but also to coherent radiation speckle so that texture assessment has limitations (74).

Tumour localization with radionuclides depends on the character of, and response to, the radiopharmaceutical in which the radionuclide is incorporated. This may be negative, i.e. the tumour may be seen as a non-radioactive zone in a radioactive background as in the case of a ^{99m}Tc-labelled colloid taken up by the liver or spleen (75), or positive as in the use of radioiodine in thyroid studies. Cold areas seen in radionuclide studies, e.g. in the liver, may be due to metastatic disease, cysts or multiple abscesses (76). Differentiation between these conditions is vital and can usually be achieved with ultrasound. The response to NMR is seen mainly through prolonged T_1 relaxation times but its specificity is not yet clear, as oedema and non-cancerous cysts also have prolonged T_1 values (49, 51, 54).

Spatial resolution in proton radiography and tomography is limited by small angle multiple scattering (77–79). The resolution improves with increasing charged particle mass but the dose also increases due to nuclear interactions (77, 79). Nevertheless, charged particle beams have considerable potential for imaging of small or low contrast tumours with lower doses than are customary for X-rays, but results for calcified metastases are inferior to those obtained with X-rays owing to the lack of contrast (61, 80, 81).

Cerebral infarctions can be recognised as zones of decreased density with little or no associated mass (57). X-ray attenuation coefficients for such a region change by a few percent whereas NMR T_1 values may change by 30–100% (54). The NMR T_1 image of brain ischaemia may also show higher contrast than that achieved by X-ray CT after injection of contrast (82). Because of the good spatial resolution achievable and the absence of risk, Budinger (54) suggests that NMR imaging may become the preferred means of diagnosis for cerebral vascular accidents. Emission CT is highly sensitive to cerebral infarction and appears more sensitive than NMR or X-ray CT to blood brain barrier disruption (54). Observations on cerebral infarction and other phenomena in the brain have been reported in studies with protons and alpha-particles, again with very low doses (80, 83, 84) compared to X-ray studies. Multiple sclerosis lesions have been observed in proton radiography (84), where they appear more transparent than the surrounding tissue, and in NMR imaging (85), but are not observed in X-ray CT.

Normal muscle is nearly homogeneous in density but when loss of muscle occurs, patchy or diffuse zones appear and the blood vessels are more prominent. These changes are clearly seen in X-ray CT without contrast. The use of topical NMR in conjunction with X-ray CT in the study of muscle metabolism is noted in section 1.5.2.

1.2.4 *Measurement of dimensions*

Quantitative measurements of foetal crown-to-rump length or of biparietal diameter provide highly accurate estimates of gestational age. These measurements must be made with ultrasound since hazards associated with other radiations are not acceptable. Precise ultrasonic methods of estimating foetal volumes are complex and time-consuming, but it is found that measurement of trunk circumference at the level of the umbilical vein is a simple and reasonably accurate method of measuring foetal weight (86).

In many clinical situations, it is necessary to estimate the size of the thyroid gland, spleen, kidney or liver, in order to determine whether their sizes are changing due to progress of disease or due to treatment. It may also be useful to compare volume data with functional data. Volumes may be deduced from ultrasound scanning (87) by multiplication of three measurements of diameter and use of a correction factor, by the use of parallel scans, or by angulated longitudinal scans.

Quantitative information which may be obtained from X-ray CT scans includes areas of surfaces, wall thickness, volumes of cavities and organs, cerebral ventricular size, and tumour volumes. Extraction of this information requires additional processing of the CT data (88). Boundary lines in successive cross-sections determine the spatial extent in three dimensions and various techniques are then used to compute the volume (89–91). Boundary definition is by no means simple and is subject to errors due to heterogeneity of tissue composition, partial volume effects (89, 92) and blurring due to internal movements (93). It is necessary to make some assumptions about the general shape of the object, because of the use of a slice-by-slice approach (88). True three-dimensional methods can handle more complex objects (94, 119).

1.3 Functional information

1.3.1 *Distribution of activity*

Emission tomography is carried out using radionuclides which emit a single photon in the radionuclide decay, or which emit a positron leading to two photon emission in the subsequent e^+e^- annihilation. Non-medical studies include the detection of active material in used nuclear fuel elements (95). In medical imaging, physiological information such as oxygen and glucose metabolism and drug response may be obtained (see Vol. 1, chapters 6 and 7). For example, in appendix A.3 of Vol. 1, models and scanning protocols are described for the regional determination of cerebral blood flow using an inert gas, cerebral oxygen metabolism using $^{15}O_2$ and $C^{15}O_2$, distribution of red cells using ^{11}CO and cerebral glucose metabolism using FDG ((^{18}F) fluoro-2-deoxy-D-glucose). Cerebral glucose metabolism has been measured in patients with focal epilepsy (96) and metabolically active regions in the brain have been visualized *in vivo* during external visual and auditory stimulations (97). FDG and $^{15}O_2$ are well established indicators of brain metabolism and energy consumption but it is thought that ^{11}C-labelled amino acids, being precursors of neurotransmitters, should also be suitable for studying brain function (98). Other important medical applications include brain blood volume studies using ^{99m}Tc-labelled red cells, brain blood perfusion using ^{123}I-iodoantipyrine, brain perfusion using ^{133}Xe or ^{127}Xe and dopamine metabolism. There is, however, need for new radiopharmaceuticals with higher specificity and higher regional uptake (99).

Budinger has concluded that statistical requirements for quantitative functional studies limit γ-ray emission tomography to objects < 20 cm in diameter (33, 37). Positron emission tomography systems offer higher sensitivity and quantitative studies are not limited to the brain and small animals; however, because the radiopharmaceuticals used often undergo extensive biochemical utilization (being labelled forms of compounds naturally abundant in the body), the compartmental models are more complex.

Functional information from fast dynamic studies can be derived using ultrasound, digital subtraction angiography, DSA, and fast X-ray CT (section 1.3.3). The growing interest in DSA is due to its ability to remove interfering effects from uninteresting structures so that the resulting image exhibits clinically significant details with enhanced visibility (100). It is possible to visualize anatomical detail as small as 1 mm at 1% contrast with a 1 mR exposure. Frame rates can vary from one image every few seconds to 30/s for fast studies. DSA can give images displaying the haemodynamic and possibly pharmacokinetic behaviour of contrast media, thereby providing physiological/functional as well as anatomical information (101).

NMR and TMR also offer potential for functional and metabolic studies (see sections 1.3.4 and 1.5.2). At 1.5*T*, both NMR imaging and chemical shift spectroscopy are possible. Hence, a whole body system with sufficient magnetic field homogeneity operating at this field strength may provide both anatomical and physiological/biochemical information.

The trends in the applications of different imaging modalities to medicine

Table 1.2 Comparison of trends of imaging methods in medicine (29).

	radionuclide imaging	X-ray CT	DSA	ultrasound	NMR
Central nervous system					
1 neoplasm	↓	↓			↑
2 carotid blood flow	↓		↑	↓	
3 cerebral blood flow	↑ IMP		↑		
4 vessel anatomy			↑		
5 aneurysm	↓		↑		↑
6 haematoma	↓	↓			↑
7 stroke	↑	↓			↑
8 oedema	?	↓			↑
9 atrophy		↓			↑
10 degenerative diseases		↓			↑
Heart					
1 infarction	↑				↑ ?
2 scar tissue					↑
3 function	↑		↑	↑	
4 muscle blood flow	↑				↑ ?
5 muscle metabolism	↑				↑ ?
6 coronary artery anatomy			↑		↑
Lung					
1 embolus	↑ antibody ?		↑		↑ ?
2 infarction	↑ IMP		↑		↑ ?
3 tumour	↑ antibody				?
4 metabolism	↑ IMP ?				?
5 function	↓		↑		
Liver					
1 cirrhosis	↓	→		↓	↑
2 tumours	↑ antibody	→		↓	↑
3 function	↑				↑
Kidney					
1. immune diseases	↑ ? antibody				↑ ?
2 inflammatory diseases		↑		↓	↑
3 metabolism					↑ ?
Bone					
1 anatomy		↑			
2 metabolism	↑				

IMP isotropyl [^{123}I] *p*-iodoamphetamine
↑ increase, ↓ decrease, → constant.

have been summarized by Deland (29), and are given in table 1.2. DSA will reduce the number of invasive angiography studies. Radionuclide antibodies will be increasingly used (section 1.1) and isopropyl (^{123}I) *p*-iodoamphetamine, IMP, will find applications in the brain and lungs.

1.3.2 *Flow measurements*

The rapid and accurate measurement of flow is of considerable importance in industry. The quantities of interest are total flow rate, velocity of flow, and also flow patterns since the mode of flow, whether lamellar or turbulent, may indicate the presence of obstructions in the flow channel. Non-invasive measurement of blood flow is of particular value in the diagnosis of major diseases (102); cerebrovascular disease affects blood movement in the brain while other vascular diseases occur primarily in the limbs, especially in the legs, and blood flow in the region of the heart reflects heart disease in various forms. The distribution of blood in an organ, a reflection of the organ's state of health, is dependent on the blood supply to the organ by major vessels, and on the state of the arterioles and capillaries; tissue perfusion is regulated by control of the arteriolar diameter (103). An artery and the corresponding vein often lie close together but can be discriminated by a method sensitive to the direction of flow.

Methods for flow studies are compared in table 1.3. Pressure methods are based on the measurement of pressure difference along the flow path. Dilution methods are based on the measurement of concentration or temperature of the diluted indicator at a site downstream from the injection (104). Measurements of concentration utilize contrast media or radionuclides; thermal methods utilize cool or warm saline. Electromagnetic methods are based on the measurement of the voltage generated when blood, a conductor, flows through a magnetic field perpendicular to the direction of flow; a wide range of velocities can be measured. Pulsed Doppler imaging, although limited to

Table 1.3 Comparison of methods for the assessment of flow.

Method	*Flow parameters measured*
volume/time methods	flow rate
pressure methods	
Pitot	velocity, velocity profile, direction of flow
Venturi	velocity
computed pressure gradient	flow rate, direction of flow
bristle, pendulum	flow rate, direction of flow
dilution methods	
concentration	flow rate
temperature	flow rate
electromagnetic methods	velocity
ultrasound methods	
pulsed transit time	velocity
continuous wave Doppler	velocity, velocity profile
pulsed Doppler imaging	velocity, velocity profile, direction of flow
NMR methods	velocity, velocity profile, direction of flow

regions accessible by ultrasound, can provide a dynamic visualization of blood flow (71).

Dynamic X-ray CT with iodinated contrast media allows the application of indicator dilution theory for the assessment of blood flow in the brain (105), the coronary arteries, coronary bypass grafts, myocardial tissue and cardiac chambers (106). Compared to angiography (107) and conventional or dynamic radionuclide methods (108), tomographic techniques such as dynamic X-ray CT, emission tomography and NMR imaging methods eliminate errors due to flow in regions adjacent to the vessel of interest and so give detailed pathological information. If blood flows with speed v in a direction perpendicular to the tomographic slice, then the blood volume in the slice will advance a distance $l = v\tau$ during the time τ. In NMR imaging using the inversion-recovery sequence, if $l > \Delta z$ (the slice thickness) and $\tau < T_1$ of blood, then the detected signal arises from blood unaffected by the 180° pulse thereby resulting in enhanced brightness compared to the case of stationary blood. If the spin-echo sequence is employed, the pixel value will be reduced by increased flow (109). In general, protons flowing through a region in which a magnetic field gradient exists will undergo changes in their resonance frequency and their phase angle; it has been shown (110) that the phase change is directly related to the velocity of flow. The flow pattern along a vessel can be studied by rapid 3-D image reconstruction in the DSR (dynamic spatial reconstructor) (111) and the CVCT scanner (cardiovascular CT) (112) using X-rays; emission tomography using γ-rays and positrons and NMR imaging can also give images of several parallel planes simultaneously (37, 113). Digital radiography with iodinated contrast media offers another method for visualizing blood flow quantitatively, through subtraction of consecutive images or temporal filtering (114).

Fast X-ray CT scanners find applications in the assessment of flow in industrial pipes. A γ-ray CT scanner utilizing 662 keV photons and divergent beam geometry centred around a pipe aims at dynamic flow imaging of gas bubbles and the study of heat transfer characteristics (115). The foaming behaviour of coal can be examined as a function of time and temperature using X-ray CT (116) by placing the furnace containing the coal within the imaging volume of the scanner.

1.3.3 *Cardiac function*

The dynamic study of the heart is the ultimate aim of fast scanners such as the DSR and the CVCT. With existing X-ray CT scanners or with scanners of scanning times ~ 0.1 s (117), physiological gating can be employed (118). Using the stored electrocardiogram (ECG) obtained during continuous projection data collection, projections are sorted according to the phases of the cardiac cycle. From the reconstructed images, the extent of myocardial contraction and ejection fractions can be derived but accuracy is limited, as in all gated studies, due to normal variations in the heart cycle from beat to beat. Because of dose considerations and the use of television cameras (with their limited dynamic range) in the fluoroscopic imaging systems, image quality and contrast are degraded and contrast media injections directly into the heart chambers via invasive catheterization may be necessary (117). However, invasive methods risk vessel wall damage, coagulation and the possibility of infection, and large amounts of high osmolality contrast media risk the possibility of adverse reactions.

Digital subtraction angiography DSA is finding applications in blood-pool and larger vessel imaging. When combined with ECG gating, cardiac ejection fractions and output data can be determined and myocardial wall motion can be assessed (14, 114).

Cardiac function can be studied quantitatively by analysing global or regional time activity curves, TAC, which reflect the variation of radioactivity with time during the heart cycle (119). Gated equilibrium measurements are made using collimated detector(s) or a γ-camera, often positioned over the left ventricle, LV. From a TAC, various systolic and diastolic measures of cardiac function can be derived such as stroke volume, ejection, filling rate and fraction of filling in early diastole. Also, since a TAC is a periodic function of time, it is amenable to Fourier analysis (see Vol. 1); the first few terms of a Fourier series are found to provide an adequate description (119). Variations in the shape of a TAC (and variations in the parameters describing it) from beat to beat can provide physiological information such as the effects of exercise and other external stimuli on cardiac function, and the immediate effects of fast acting drugs (120). When time activity curves are simultaneously measured on a pixel by pixel basis, a series of images can be evaluated (using the stored TAC data) each one representing the map of a different cardiac parameter over the field of view or tomographic plane (121). Because of the shape of the left ventricle, use of a polar imaging grid can be advantageous in some cases (119). The usefulness of global and regional parameters describing cardiac function rests on the statistical reliability of the TAC data. Hence the use of an array of high efficiency detectors spanning the region of interest would be preferable to a NaI γ-camera.

As noted in section 1.2.1, ultrasound is used in cardiac imaging, by means of the two complementary techniques of real-time scanning and M-mode echocardiography. As table 1.2 indicates, NMR appears promising for heart studies but its usefulness has not been established yet. Whether with X-rays, γ-rays, positrons, ultrasound or NMR, cardiac imaging will ultimately improve the diagnosis and management of heart disease and will influence the surgical and pharmacological regimes.

1.3.4 *Follow-up of disease and therapy*

Because NMR imaging is believed to be non-hazardous and can provide different images of the same anatomical slice reflecting different tissue properties (both physical and biochemical), it is hoped that it may offer greater potential than other imaging methods for the follow-up of disease and the effectiveness of therapy. With the exception of radio-therapy, changes incurred during therapy regimes of physical properties such as density, proton density, attenuation coefficients, stopping power and acoustic refractive index, are likely to be small but changes in the biochemical environment may be substantial.

Changes in metabolic processes due to the presence of disease are known to cause changes in the abundance of minor and trace elements. Of these, iron and copper are paramagnetic and increased concentrations reduce the spin-lattice relaxation time T_1, as in haemosiderosis (a liver disease) where the liver iron content increases (122). Although both liver iron content in haemochromatosis and liver copper in cirrhosis increase, the T_1 is also dependent on other tissue reactions and a direct relationship between T_1 and concentration does not exist. In liver disease, NMR compares favourably to X-ray CT in indicating the nature of lesions (123), is as sensitive as ultrasound for the demonstration of space-occupying lesions and more sensitive than radionuclide imaging for metastases less than 15 mm in diameter (122).

High resolution ^{31}P NMR can provide regional biochemical information that can be related to the onset of disease and the monitoring of therapy (204). During stroke induced in a monkey, different biochemical changes were observed in two different regions of the brain. During shock induced in a rat, the time-course of changes in kidney metabolites was obtained and the kidney was shown to be the first organ that biochemically fails in a shock situation. In humans, changes in muscle metabolism can be observed during stress and exercise. By detecting the onset of disease through changes in the TMR spectra, it may be possible to begin a therapy regime before the disease causes any irreversible damage. Similarly, the effectiveness of the regime can be monitored.

The difference in elemental composition between grey and white matter within the brain leads to their differentiation using X-ray CT scanners, but the difference is small (124). Although the elemental composition is similar, the chemical environment differs substantially in the two tissues: there is more hydrogen in the form of water in grey matter than in white matter where a higher proportion is bound in fats and large molecules. This leads to their striking differentiation when NMR imaging methods are used that depend on chemical environment (and hence on T_1), such as the inversion recovery sequence $180^\circ - \tau - 90^\circ$ (82; chapter 4). Hence, small differences in the relative degrees of grey and white matter involvement during a disease process may be discernible by NMR and this suggests a potential for the follow-up of the progress of the disease and for monitoring the effectiveness of therapeutic regimes. There are infective, metabolic, toxic and nutritional diseases in which demyelination is a prominent feature; their time course could also be assessed by NMR. For example, in multiple sclerosis abnormalities have been revealed on a scale not previously seen except at necropsy (85).

1.4 Analysis

1.4.1 *Density*

For a single element with n atoms present per unit volume, the relation between the physical density ρ and the electron density nZ, the number of electrons per unit volume, is given by

$$nZ = \rho N_g$$

where $N_g = N_A Z/A$ is the number of electrons per unit mass, N_A is Avogadro's number and A is the atomic weight. For a material composed of a mixture of elements, this relation becomes

$$\sum_i n_i Z_i = \rho \sum_i N_g^i = (\rho N_g)_M \sum \lambda_i \tag{1.32}$$

where

$$N_g = \sum_i N_g^i = N_A \sum_i (\omega_i Z_i/A_i) \tag{1.33}$$

$$\lambda_i = (\omega_i Z_i/A_i)/\sum_j (\omega_j Z_j/A_j) \tag{1.34}$$

and ω_i is the proportion by weight of the ith element. In all the physical measurements discussed in this section, the electron density is the quantity directly related to the measurement and the physical density is inferred from it. Table 1.4 gives values of physical density relative to water $\rho_r = \rho_M/\rho_W$ and electron density relative to water $E_r = (\rho N_g)_M/(\rho N_g)_W$ for a wide range of tissues and tissue substitutes.

In charged particle radiography and tomography (Vol. 1, chapter 5) the principal cause of energy loss is ionization. The rate of energy loss (stopping power) is given by the Bethe-Bloch formula

$$S_M = -\frac{dE}{dx} = -(\rho N_g)_M \frac{KZ_p^2}{\beta^2}\left[\log_e \frac{K'\beta^2}{I_M(1-\beta^2)} - 2\beta^2\right] \tag{1.35}$$

where Z_p is the charge of the projectile, $v = \beta c$ is its velocity, K and K' are

Table 1.4 The calculated values of stopping power and CT number (127) for some materials, tissue substitutes and tissues (128).

Material	*Calculated relative electron density*	*Calculated relative stopping power*	*Reconstruction field size* mm	*Calculated relative attenuation coefficient*		*Calculated CT number (Hounsfield units)*	
				120 kVp spectrum	140 kVp spectrum	120 kVp spectrum	140 kVp spectrum
Water	1.00	1.00	200	1.00	1.00	0.00	0.00
			320	1.00	1.00	0.00	0.00
Aluminium	2.35	2.12	200	3.07	2.936	2065	1936
			320	2.91	2.79	1907	1786
Lucite	1.16	1.140	200	1.13	1.133	130	133
			320	1.134	1.137	134	137
Polyethylene	0.947	0.970	200	0.91	0.915	−91	−85
			320	0.916	0.922	−83	−78
Polystyrene	1.052	1.061	200	0.99	0.989	−10	−11
			320	1.030	1.034	30	35
Paraffin (wax)	0.932	0.954	200	0.894	0.90	−106	−102
			320	0.902	0.907	−98	−93
Polyurethane + Al_2O_3	0.278	0.264	200	0.282	0.279	−718	−720
			300	0.279	0.277	−720	−722
Nylon	1.130	1.143	200	1.105	1.11	105	110
			320	1.111	1.115	111	115
Teflon	1.892	1.79	200	1.97	1.95	968	955
			320	1.952	1.940	953	940
Cysteine	0.334	0.325	200	0.394	0.38	−606	−616
			320	0.381	0.372	−619	−628
Benzene	0.855	0.861	200	0.830	0.834	−170	−166
			320	0.835	0.839	−165	−161
Ethyl alcohol	0.806	0.812	200	0.784	0.787	−215	−213
			320	0.788	0.791	−212	−208
Heptane	0.746	0.767	200	0.714	0.718	−286	−281
			320	0.720	0.724	−280	−276
Propene	0.536	0.548	200	0.514	0.517	−486	−482
			320	0.518	0.521	−482	−479
Glycerol	1.238	1.22	200	1.22	1.22	224	226
			320	1.23	1.23	227	229
Ethyne	0.600	0.605	200	0.583	0.586	−417	−414
			320	0.587	0.590	−413	−410
Muscle	1.054	1.054	200	1.057	1.056	57	56
			320	1.056	1.055	56	55
Fat	0.931	0.946	200	0.903	0.904	−97	−95
			320	0.907	0.911	−93	−89
Lung	0.25	0.288	200	0.281	0.279	−719	−720
			320	0.279	0.278	−720	−722

Table 1.4 (*contd.*)

Material	*Calculated relative electron density*	*Calculated relative stopping power*	*Reconstruction field size* mm	*Calculated relative attenuation coefficient*		*Calculated CT number (Hounsfield units)*	
				120 kVp spectrum	140 kVp spectrum	120 kVp spectrum	140 kVp spectrum
Soft bone	1.1	1.08	200	1.332	1.282	333	282
			320	1.272	1.220	273	224
Hard bone	1.74	1.645	200	3.211	2.96	2211	1964
			320	2.917	2.682	1917	1682
Compact bone	1.634	1.59	200	2.138	2.06	1138	1056
			320	2.04	1.96	1041	963
Total skeleton	1.311	1.278	200	1.783	1.69	783	693
			320	1.676	1.59	676	591
Lung 2	0.407	0.404	200	0.380	0.378	− 620	− 622
			320	0.378	0.377	− 622	− 623

constants, and I_M is the mean ionization energy. (The calculation of I_M is discussed in section 1.5.1). Unnecessary correction terms have been omitted. Because the quantity in square brackets in eq. (1.35) varies only slowly with energy, the stopping power relative to water $S_r = S_M/S_W$ is, to a good approximation (125), a constant which is independent of the nature of the charged particle and depends primarily on the relative electron density E_r (126). Calculated values of S_r and E_r are given in table 1.3 for some materials, tissue-substitutes and tissues (127). The values S_r are plotted against E_r in figure 1.2. These results confirm that there is a linear relationship between relative stopping power and relative electron density.

It follows from eq. (1.35) that changes in the electron density of the material will be reflected in variations in the residual energy or residual range of the particles after they have passed through the material. By considering an inclusion of material of density ρ_I in a matrix of density ρ_M, it is possible to calculate the thickness Δt_M (mass per unit area) of matrix material which would have the same effect as the inclusion of thickness Δt_I. This is given by (128).

$$\Delta t_M = \Delta t_I\,[S_r \rho_I/\rho_M) - 1]. \tag{1.36}$$

Eq. (1.36) implies that the radiographic contrast produced by the inclusion is independent of its position within the matrix, and hence that the sample can be placed in that part of the path where the energy loss (dose) and multiple scattering are low (130). The effect of irregularities in the shape of the sample can be suppressed by surrounding it with a reference material R for which $S_R \rho_R / S_M \rho_M = 1$ (129).

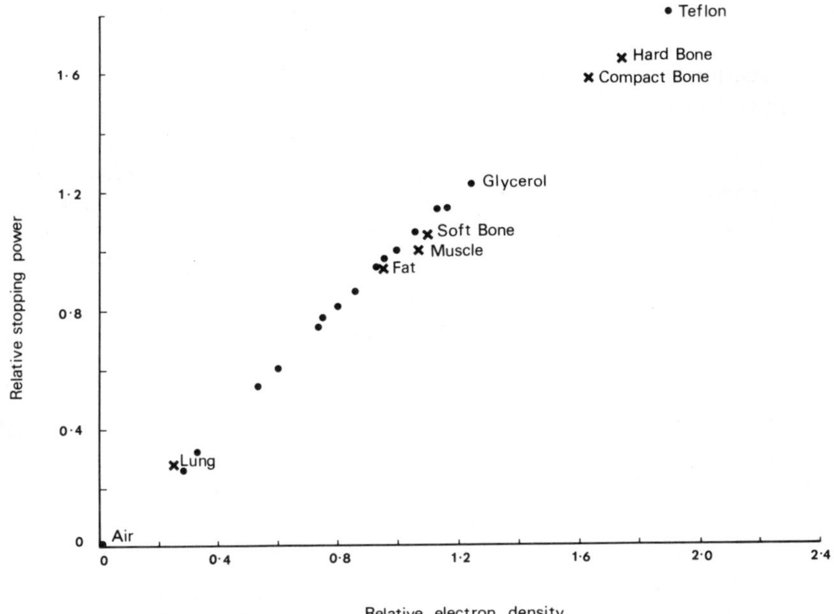

Figure 1.2 Variation of stopping power relative to water with electron density relative to water (127). (The crosses denote values for tissue while the filled circles denote values for various chemical compounds listed in table 1.4).

The criterion for detectability by eye of discontinuities in a homogeneous background is that the signal-to-noise ratio (SNR) should exceed 5 (131). Hence, for detectability

$$\Delta t_M \geq 5\delta t \tag{1.37}$$

where δt is the uncertainty in the measurement, which arises mainly from the range straggling of the beam (81). For 200 MeV protons with a range R of 280 mm in water, the fractional straggling is $\sigma_R/R = 0.0106$ and hence a change in thickness of 0.053%, or an equivalent change in density over a fixed length, can be detected (132). The detectability increases with energy and with increasing projectile mass; for energies chosen to give the same range in water, $\Delta t_M/R_M$ is 0.027% for ^{4}He ions, 0.014 for ^{12}C ions and 0.011% for ^{20}Ne ions (81, 132). These values can be achieved in practice (61, 81, 129, 133). Thus ^{12}C ions are about 3.75 times better for resolution than protons over the same range, but this is achieved at the cost of increasing the particle energy and increasing the dose at the centre of the sample by a factor of 1.53 (132, 134).

A number of slightly different formulae have been given for the mean square scattering angle arising from multiple Coulomb scattering (79, 135). They have the common feature that $\theta_{\mathrm{rms}} \propto (\rho N_g Z)^{1/2}$ for different materials of the same thickness $t=\rho L$ and $\theta_{\mathrm{rms}} \propto (\rho L)^{1/2}$ for different thicknesses of the same material. It has been noted that if two high energy particles traverse two adjacent regions, causing different rms scattering angles, the resulting intensity recorded on a suitable detector will show a modulation in the vicinity of the boundary (136). This technique picks out the edges of internal structures in the presence of various thicknesses of overlying materials. Some experiments with 160 MeV protons have been carried out (136, 137).

In nuclear scattering radiography (Vol. 1, chapter 5), the flux N_s of high energy protons scattered by a volume element V is given by

$$N_s = 0.6 N_i V \sigma \rho / A \tag{1.38}$$

where N_i is the incident flux per unit area and σ is the integral of the relevant differential cross-section over the solid angle of the detectors looking at the scattered particles. Measurements of the scattered flux with two position sensitive detectors make possible the imaging of density variations. The main limitations of the method are high dose rates and slow data acquisition times. Emphasis to date has been on the potentiality of the method for medical imaging (62, 138) but for non-medical objects, when dose and time are less restrictive, quite high accuracy for density variations might be achievable.

Because of the complexity of the photon-atom interaction it is only when the photon energy is such that the incoherent scattering process dominates and can be accurately represented by the Klein-Nishina formula that simple X-ray measurements can give a direct measure of the electron density. In this case, the photon attenuation coefficient can be written in the form

$$\mu_M(E) \simeq (\rho N g)_{Me} \sigma^{KN}(E) \tag{1.39}$$

where σ^{KN} is the Klein-Nishina cross-section for scattering from a single, free electron. This situation is not reached until $E \sim 250\,\mathrm{keV}$ for light elements and $E \gtrsim 400\,\mathrm{keV}$ for high Z elements (12). Above $E = 2mc^2$, the pair production process also contributes and the attenuation coefficient is no longer proportional to $(\rho N_g)_M$. Except at very low energies, the coherent scattering is strongly forward peaked, due to the presence in the differential cross-section of the square of the form factor $F_0(q)$, where $\hbar q$ is the momentum transfer, while the incoherent scattering disappears at small angles owing to the presence in the differential cross-section of the incoherent scattering function $S(q)$. At sufficiently large angles, the incoherent scattering is dominant.

Methods have been developed for direct measurement of electron density, and these have recently been reviewed (139). In scattering densitometry, a fixed point is sampled, whereas in scattering tomography a series of points are sampled to give a tomographic image. Doubts have been expressed about the tomographic methods for medical applications due to the high dose involved. Results of electron density measurements for non-biological materials are encouraging (140). An accuracy of 4.3% in the measurement of electron density is achieved with the principal uncertainties arising from multiple scattering and corrections for attenuation and background detection.

1.4.2 *Elemental composition*

A very large number of techniques are available for investigation of elemental composition. In figure 1.1, we distinguished between those investigative techniques which are primarily diagnostic and those which are primarily analytical. This distinction is somewhat arbitrary in the sense that there is no sharp boundary, for example, between microradiography and microscopy. When the determination of elemental composition is considered in detail, however, there is a very important practical distinction between those techniques which are applicable only to surface studies or to sections of severely limited thickness and those techniques which give information about the average composition of matter in bulk or about localized regions within a larger specimen. For those who are concerned with medical imaging and non-destructive testing, the latter group of techniques is of principal interest. However, the increasing availability of intense, tunable, photon beams at synchrotron radiation sources has aroused considerable interest in different aspects of imaging, particularly of the microstructure of biological materials. These applications have been reviewed extensively (141) and we have discussed elsewhere (139) some of the attractive features and potential limitations of the use of soft X-rays.

X-ray fluorescence and particle-induced X-ray emission are important techniques for high precision trace element analysis. A very complete review of PIXE has recently appeared (142), which gives comparisons with XRF and with induced X-ray emission by other particles. An interesting application of PIXE to the study of calcium distribution in pollen grains in various states of development has recently been reported (143). XRF is used for analysis *in vivo* in man to measure, in various locations, strontium (144), iodine (145), cadmium (146) and lead (147).

Materials analysis by X-ray transmission measurements has recently been reviewed (139; Vol. 1, chapter 4). Measurements at one photon energy are useful for detection of one known component in a known matrix and in this case there is no advantage in a dual energy measurement. The results can be represented in terms of a detectable mass fraction (148), or a detectable length fraction which is simply related to the Hounsfield number for the added component compared with the matrix as reference material (149). The optimum energy is normally the one at which the mass or length fraction has its minimum value. Accuracy is highest for detection of a heavy element (high Z) in a light matrix (low Z) when accuracies of 1 part in 10^4 can be achieved.

Measurement of the ratio of elastic (coherent) to inelastic (incoherent) X-ray scattering has been investigated as a method of determining bone mineral content (150). The optimum energy is quoted as around 80 keV, when the error is $\lesssim 0.9\%$ for a tissue layer at 10 cm surrounding human bone. This type of measurements has also been explored by Kerr *et al.* (151) who proposed a measurement of coherent scattering alone which should be more sensitive to changes in mineral composition. Conventional CT scanners are also used for bone densitometry (152, 153) but this method is adversely affected by beam hardening effects. Special purpose CT systems have been developed (153, 154) using a radionuclide source, usually ^{125}I which provides a low energy monochromatic beam, and good discrimination between soft tissue and bone is obtained. Doses as low as 3 mrad are possible but the low energy of the beam limits the size of objects which can be studied.

The elastic scattering of charged particles has been developed as a tool for analysis, as both yield and scattering angle depend on the composition of the sample (155). Rutherford backscattering (RBS) is a quantitative non-destructive method used to study surface layers and to calibrate other methods. Hyvönen-Dabek *et al.* (156) have applied this method to bone samples for which the RBS spectrum shows a number of separate steps whose heights are proportional to the abundance of one of the elements. The analysis requires a knowledge of the stopping powers and a prior knowledge of the approximate composition of the sample. It is claimed that this method is more accurate than neutron activation analysis and gives reliable results for the Ca/O and P/O ratios as well as Ca/P.

The activation of a sample by bombardment with charged particles, neutrons and γ-rays produces a rich source of analytical information. A comparison (157) of neutron and proton activation of biological samples has indicated that, with delayed photon counting, neutron activation determines a greater number of elements with high sensitivity but proton activation indicates the presence of certain elements not accessible to neutrons, e.g. titanium and strontium. Charged particle activation is used very effectively to determine concentration of light elements, especially hydrogen and oxygen. When combined with detection of X-rays from radionuclides this method gives a means of non-destructive multi-element trace analysis (158). Heavy ion activation analysis can also be combined with detection of elements with both high and low Z (159). This method has been applied to metalloproteins and can be applied to microsamples.

Table 1.5 Measurement of body elements by *in vivo* neutron activation analysis (160).

Stable element	*Amount in 70 kg reference man* (g)*	*Proportion by weight 70 kg reference man* (%)*	*Induced nuclide*	*Neutron reaction*	*Gamma or X-ray emission*
Oxygen	43 000	61	^{16}N	n, p(fast)	delayed γ(6–7 MeV)
Hydrogen	7 000	10	^{2}H	n, γ(thermal)	prompt γ(2.2 MeV)
Nitrogen	1 800	2.6	^{13}N	n, 2n(14 MeV)	delayed γ(0.51 MeV)
				n, γ(thermal)	prompt γ(10.8 MeV)
Calcium	1 000	1.4	^{49}Ca	n, γ(thermal)	prompt γ(many); delayed (3.10 MeV)
			^{37}Ar	n, α(14 MeV)	delayed X-ray (2.6 MeV)
Phosphorus	780	1.1	^{28}Al	$n. \alpha$(fast)	delayed π(1.78 MeV)
			^{32}P	n, γ(thermal)	prompt γ(0.08 MeV)
Sodium	100	0.14	^{24}Na	n, γ(thermal)	prompt γ(many); delayed (2.75 MeV)
Chlorine	95	0.12	^{38}Cl	n, γ(thermal)	prompt γ(many); delayed (1.6 and 2.2 MeV)
			^{37}S	n, p(fast)	delayed γ(3.10 MeV)
Magnesium	19	0.03	^{27}Mg	n, γ(thermal)	delayed γ(0.84 MeV)
			^{24}Na	n, p(fast)	delayed γ(2.75 MeV)
Iodine	0.1	< 0.01	^{128}I	n, γ(thermal)	delayed γ(0.45 MeV)
Cadmium	Trace	Trace	^{114}Ca	n, γ(thermal)	prompt γ(0.559 MeV)

*Determination by chemical analysis of human cadavers, ICRP 23, 1975.

Measurement *in vivo* of various elements by neutron activation analysis has recently been reviewed (160). The elements which can be determined by this means are listed in table 1.5. Other elements, such as selenium, may be determined by cyclic activation analysis (161). Partial body calcium has also been measured by proton activation analysis (162). Other methods adopted are the use of photon activation via the (γ, n) reaction leading to positron emitters for bulk analysis of carbon, nitrogen and oxygen in regions of the body (163), measurement of whole body carbon by detection of de-excitation γ-rays following the $^{12}C(n, n')$ reaction (164) and use of nuclear resonant scattering of γ-rays for measurement of copper (165) and iron (166).

Nuclear scattering radiography (Vol. 1, chapter 5) can be used to image hydrogen separately from other elements, but incident nucleons in the energy range 500 MeV—1 Gev are required. Neutron radiography depends on the neutron cross-section which is strongly dependent on isotopic composition and neutron energy (167). Neutron attenuation coefficients for certain light elements such as H, Li, Be are very high compared with many heavier elements. (The high attenuation for hydrogen comes from the scattering cross-section and this presents difficulties with thick targets of hydrogenous material). These properties have the effect that neutrons are rather easily transmitted through bone so that neutron radiography can be used to examine soft tissue which may be masked by bone (168). The feasibility of neutron tomography is under investigation (169); this could be an effective means of imaging selected elements, particularly hydrogen.

It is significant that, of the important elements mentioned in this section, only hydrogen can be imaged directly on a macroscopic scale, without the use of contrast agents or multi-energy measurements. This can be done by nuclear scattering radiography and neutron radiography and tomography, both of which are sensitive to all hydrogen present, and by NMR imaging which is sensitive to the hydrogen in free water. For this reason, there has been considerable interest in dual- and multi-energy methods. By using scanning microscopy and similar techniques, many elements may be mapped (i.e. imaged) on a microscopic scale.

1.4.3 *Dual-energy and multi-energy studies*

Because the contributions to the photon attenuation coefficient in the energy range 10–1000 keV come from three separate physical processes (coherent scattering, incoherent scattering and the photoelectric effect) which show different dependence on photon energy E and atomic number Z, determination of quantitative information with photon beams is generally more complicated than for neutrons or charged particles. Exceptions arise only in special circumstances when one of the three processes dominates, as described

Table 1.6 The Hounsfield coefficients H_c and H_p of a calcium chloride solution calculated at 80 keV and derived from Brooks dual energy method (183).

	Calculated at 80 keV	*Derived from dual energy analysis*	*Experimental error %*	*Discrepancy between calculation and experiment %*
H_c	122.1	108.7	± 1.4	11.0
H_p	2652.8	2838.1	± 0.8	-7.0

in sections 1.4.1 and 1.4.2. A further practical complication arises through the difficulty of obtaining sufficiently intense monochromatic beams in most laboratories and clinical centres. For non-medical applications, this difficulty may be overcome by the use of specially selected isotopic sources or by use of radiation from synchrotron sources, but for medical applications the X-ray tube is likely to remain the principal source of photons.

The effect of a polychromatic beam can be taken into account by averaging the linear attenuation coefficient over the X-ray spectrum (Vol. 1, chapter 4). This procedure yields the relative attenuation coefficients and CT numbers listed in table 1.4. When plotted against relative electron density E_r, the averaged attenuation coefficients show linearity up to $E_r \sim 1$, as shown in

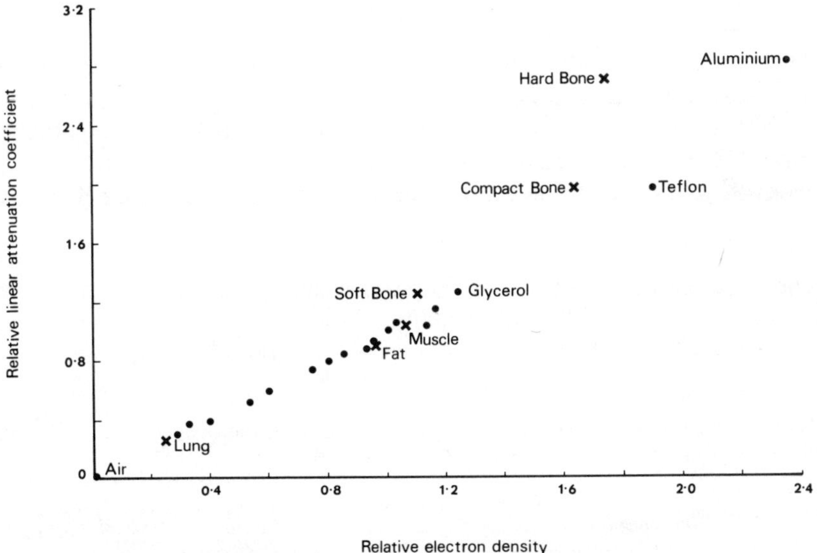

Figure 1.3 Variation of averaged attenuation coefficient relative to water with electron density relative to water (127). (The crosses denote values for tissue while the filled circles denote values for various chemical compounds listed in table 1.4).

figure 1.3. The departure from linearity for $E_r > 1$ arises from the strong photoelectric contribution from high Z elements.

The determination of quantitative information about a complex material from multi-energy studies requires a representation of the response of the material in terms of two or more energy-independent constants. This has been attempted by three methods—representation in terms of different processes contributing to the attenuation coefficient (170–172), representation in terms of properties derived from the attenuation coefficient (173, 174), and representation in terms of two or more components of the material (12, 148–149, 175–181).

The first of these methods depends on the assumption that the linear attenuation coefficient can be represented as the sum of the true Compton scattering, whose energy-dependence is given by the Klein-Nishina formula, and another term which represents the photo-electric and coherent scattering contributions plus binding corrections to the Compton term *and that* the energy dependence of these terms is common to all elements. We have already shown (12; Vol. 1, chapter 4) that this method is fundamentally incorrect and leads to inaccuracies in the predicted attenuation coefficients in excess of experimental uncertainties. Some further results are given in table 1.6, where the Hounsfield Compton coefficient H_c and photoelectric coefficient H_p (171) for a known $CaCl_2$ solution have been calculated exactly at 80 keV using the tabulated EMI spectrum (182) and have also been derived by a dual energy analysis (183). It is clear that the discrepancy between the calculated and derived values is well in excess of the uncertainty in the experimental measurements.

In the second approach, the material is represented by a hypothetical material with atomic number Z^* and electron density n^*Z^*, such that the linear attenuation coefficient in the region of interest is given by (173)

$$\mu_M(E) = n_a^* \sigma(Z^*, E).$$

Hence, two measurements at different energies should determine the two parameters n^*Z^* and Z^* which characterize the material. So far, this approach has been applied with rather inaccurate representations of ${}_a\sigma(Z^*, E)$, but in

Table 1.7 NMR properties of some important nuclei (205). The range quoted for the chemical shifts corresponds to those species commonly found in biological systems.

Nucleus	*Abundance*	*Relative sensitivity*	*Resonant frequency* at 1.89T (MHz)	*Range of chemical shift* (ppm)
^{1}H	99.99	1	80.3	10
^{13}C	1.11	1.8×10^{-4}	20.2	200
^{31}P	100.00	6.6×10^{-2}	32.5	40

principle it is capable of a variety of applications, e.g. assessment of brain lesions (173), investigation of iron overload in haemochromatosis (174, 184), of fatty infiltration of liver (174), and vertebral bone densitometry (174).

The third and most satisfactory approach involves representation of the linear or mass attenuation coefficient of a material in terms of separate components. The most suitable representation depends on the application and method of analysis, it may be in terms of a single element in a known matrix (148, 149, 175, 176), of groups of elements of similar atomic number (12), or of reference materials which may, or may not, be part of the actual material under study (12, 178, 180–181). (Although Riederer *et al.* (180) suggest that they are following the first method described above this is not the case as their representations for different materials have different coefficients and hence have different energy dependence).

It appears that soft tissues, bone, and contrast material in normal concentrations can be represented very accurately by only two components (see Vol. 1, table 4.9). This observation has some important consequences. Firstly, the contrast achieved by subtraction of two images taken at different energies is independent of the choice of the two energies provided both are above the *K*-edge of the heaviest element present (179). Secondly, representation in terms of a two-component system can lead to mis-estimates in more complicated situations. In Vol. 1, chapter 4, we discussed the case of muscle, represented by water and mineral (i.e. bone), and showed that the additional presence of fat, also represented by water and mineral with the coefficient of the latter component being negative, would lead to apparent cancellation of some of the mineral content. Riederer *et al.* (179) give a similar example of a system consisting of lucite, bone and iodine in water and show that because water can be represented by lucite and bone it mimics the presence of bone. This effect can be resolved only by using three beams of different energies.

The formalisms for applications of this last method, including the determination of errors and definition of minimum detectable quantities, have been given elsewhere (Vol. 1, chapter 4; 139, 149, 170, 177). In tomography, there is a practical difficulty arising from the effect of patient movement since the two scans are taken independently. In order to overcome this difficulty, the use of a split detector system has been proposed (185). This makes it possible to take dual energy images in one scan and also produces some chemical discrimination. Alternatively, modification of the detector collimator has been investigated (172), with metal foils covering alternate slits. The projection data then consist of two interleaved sets and are separated using software. Filters of iodine, cerium and brass have been used to give two beams which straddle the iodine edge at 33 keV and one at a slightly higher mean energy (177).

Atkins *et al.* (175, 176, 186) have developed a transmission rectilinear scanning device using fluorescent X-rays from a secondary target, high resolution detectors and tight collimation. This yields spatial resolution of ~ 3 mm and energy acceptance of ± 1.25 keV (176). Scans are taken on either side of the iodine edge to study human thyroid and either side of the calcium edge for thin bone samples *in vitro*. Two and three beam techniques have been developed by Riederer *et al.* (180, 181) using an image-intensifier fluoroscopy system. Emphasis is on the K-edge in iodine, and iodine, cerium and brass filters are again used. Kruger *et al.* (178) have developed a film subtraction technique for detection of calcification in solitary pulmonary nodules. This method is less sensitive than fluoroscopy and dual energy CT but is simpler and involves much lower levels of patient exposure. Kirz (187) has studied soft X-ray methods and has compared the dose delivered to biological specimens in dual energy absorption measurements and X-ray fluorescence measurements. This method is not highly sensitive to trace elements and requires concentrations of $\sim 1\%$ by weight.

1.5 Sensitivity to molecular structure and chemical environment

1.5.1 *Influence on the interpretation of imaging data*

The objective of this section is to consider whether the physical measurements on which the various imaging methods are based are, under any circumstances, sensitive to molecular structure and chemical environment. If such sensitivity occurs, it is necessary to know whether this may influence adversely the interpretation of imaging data or whether the extra sensitivity provides new information which may, for example, assist in differentiation between normal and abnormal tissue.

In charged particle studies, the influence of molecular effects would appear in the evaluation of stopping powers for materials. The ionization energy is calculated using the *Bragg additivity* rule, i.e.

$$\log_e I_M = \sum_i (\omega_i Z_i/A_i) \log_e I_i / \sum_i (\omega_i Z_i/A_i) \tag{1.40}$$

where Z_i, A_i are the atomic number and atomic weight of the ith element, ω_i is the proportion by weight and I_i is the ionization energy. Steward (184) has suggested that the influence of departures from this rule on stopping powers could provide new information which would be complementary to that obtained from departures from the mixture rule for X-ray attenuation coefficients (see below) because in the latter information may be less readily available for light elements. More recent data on stopping powers than those considered by Steward suggest (188) that molecular effects are not as large as previously assumed but that significant differences are observed for organic molecules in different physical phases. This latter effect has an important influence on dosimetric measurements but not normally on imaging data.

In X-ray transmission studies, detailed analysis of the data depends on the assumption that contributions to the attenuation coefficient are additive. This yields the so-called *mixture rule* for the mass attenuation coefficient of a material consisting of a mixture of elements, i.e.

$$[\mu/\rho]_M = \sum_i (\mu/\rho)_i \tag{1.41}$$

so that the linear attenuation coefficient has the form previously used,

$$\mu_M = (\rho N_g)_M \sum_i \lambda_{ie}\sigma_i \tag{1.42}$$

The underlying assumption here is that the atomic wave function of each element is not significantly modified by the molecular bonding and chemical environment, at least in regions which give the major contribution to the matrix elements for photon-atom interactions. The limits of this assumption are still not well established but it has been estimated (189) that for validity of the mixture rule the photon energy should not be below 10 keV or less than 1 keV from a photoelectric absorption edge.

The actual energy at which a particular photoelectric absorption edge of an element occurs is dependent on the chemical state of the absorbing atom and the nature of the chemical environment (190). High resolution studies show that within the edge region there is structure which depends on the electronic structure; for example there are significant differences between the satellite peaks in the 1*s* photoelectron spectra of carbon and oxygen atoms in the same molecule (191). The actual height of the photoelectric edge is also sensitive to the local atomic environment and hence there may be uncertainty in determining the *K*-jump ratios (192). On the high energy side of an edge, small amplitude oscillations are seen up to a few hundred eV above the edge, even for low *Z* atoms (193). This extended X-ray absorption fine structure (EXAFS) arises from interference between the outgoing wave representing the electron emitted from the primary atom and the reflected waves from other atoms in the material (194). This process can be described adequately by a single scattering analysis and it is now customary to distinguish EXAFS from the more complex X-ray absorption near edge structure (XANES) which covers the first 50 eV above the edge and requires a more complicated multiple scattering analysis (195).

It may be concluded that the validity of the mixture rule is very uncertain in the soft X-ray region. The lower limit for proximity to an absorption edge may be nearer 500 eV rather than 1 keV, but below this limit high resolution studies can be expected to show departures from the mixture rule of a few percent. This discrepancy is particularly relevant to imaging methods which involve subtraction of data taken in the vicinity of a K-edge.

Some calculations of the photoelectric cross-section (12, 196–198) rely on an approximate treatment of the screening of the nucleus by the atomic electrons which predicts that the effects of screening can be represented by a factor that takes account of renormalization of the wavefunction of the bound electron and a second factor which takes account of renormalization of the wavefunction of the continuum electron and the change of shape of both wavefunctions. At photon energies very much greater than the K-edge the second correction is negligible, but at energies near to that of the edge the corrections are of the same order of magnitude (198). Many calculations of the coherent cross-section (12, 196, 199) rely on the form factor approximation for the total cross-section,

$$ {}_{a}\sigma^{coh} = \int F_0^2(q) \frac{d}{d\Omega} {}_{e}\sigma^{Th} \, d\Omega, $$

where $d_e\sigma^{Th}/d\Omega$ is the Thomson cross-section for scattering from a single electron, $\hbar q = 2(E/mc^2)(mc)\sin\frac{1}{2}\theta$ is the momentum transfer and the atomic form factor $F_0(q)$ is the Fourier transform of the electron density distribution which depends on the wavefunction of the atom in its ground state. It has been shown (200) that this approximation is valid when $Z/137 \ll 1$, $\hbar q \ll mc$, and $E \gtrsim 2\varepsilon$, where ε is the electron binding energy. Thus, when attenuation coefficients are taken from tables or obtained from standard codes, it is necessary to be aware of the approximations used in the calculations if the tabulated values are to be used for comparison with data taken near to an edge.

1.5.2 *Chemical shifts*

The occurrence of a chemical shift in the energy of absorbed or emitted radiation is an important source of information about the chemical environment. For example, the actual energy of the photo-electric absorption edge of an element depends on the oxidation state and nature of the immediate chemical environment of that element (190). The satellite pattern observed in induced X-ray emission shows a chemical shift which is also dependent on the chemical environment (201) and arises from the relaxation of outer shells prior to inner shell decay. Total yields are also sensitive to the chemical environment (202). The ionization energies observed in photoelectron spectroscopy also show a chemical shift, of the order of a few eV, due to changes in the total electronic charge of the emitting atom. It is possible to compare the effect of the presence of the same element in different compounds to distinguish inequivalent atoms of the same type in a molecule or to distinguish between atoms of the same element on non-equivalent lattice sites (203).

One of the most important phenomena to show a chemical shift is nuclear magnetic resonance. In the conventional NMR spectrum, as used in chemistry and biochemistry, the position of each line associated with a particular nucleus depends on the chemical environment. The chemical shift δ is obtained by reference to a standard, through the expression

$$\delta = \frac{\nu(\text{sample}) - \nu(\text{standard})}{\nu(\text{standard})} \times 10^6.$$

It is thus a dimensionless quantity and is quoted in ppm. Similar nuclei in similar environments have the same chemical shift so that the strength of the line measures the amount of material present.

Values of the range of the chemical shift for some important nuclei occurring in biological materials are given in table 1.7. The range of chemical shifts associated with ^1H is small and the signal from biological materials is dominated by that from water. For this reason, the emphasis so far has been on studies of ^{31}P and, more recently, on ^{13}C. These studies are carried out

Table 1.8. Applications of ^{31}P NMR in the topical magnetic resonance method (204).

Tissues	*Investigations*	
Muscle	Contraction and recovery	
	Fatigue and energetics	
	Regional ischaemia	
	Dystrophy	
Heart	PCr ATP interconversion rates	
	Ischaemia	myocardial infarction
		acidosis (pH)
		drug therapy
		atheroma
Kidney	Ischaemia	
	'Transplantability'	
Liver	Ischaemia	
	Drug intoxication	
Brain	Ischaemia—cerebral infarction	
	Clinical brain death	
	Coma	
	Dementia	

by the TMR technique (chapter 4, 204, 205) which gives high resolution spectra from a small localized volume within a larger specimen. No attempt is made to reconstruct an image.

The chemical shifts depend on many parameters, including pH, cell structure, tissue type, and temperature. Hence measurements of chemical shifts can give information on cellular and organ metabolism which indicate the physiological state of living tissues. A list of applications using ^{31}P is given in table 1.8. Key features are the measurement of intracellular pH and of the time variation of metabolite concentrations which can provide information on tissue ischaemia (206), on diseases affecting muscle metabolism (207, 208) and on the metabolism of tumours (209). Signals from naturally occurring ^{13}C have been obtained (205) and arise primarily from triglycerides in fatty tissue. The ^{1}H signals from fat arise from the same source, and comparison of fat and water signals can indicate fatty infiltration into muscle which occurs in muscular dystrophy (208). (This study also made use of ^{31}P NMR, and used X-ray CT to give an independent estimate of tissue composition). It is likely that development of ^{13}C NMR will require the use of ^{13}C-labelled metabolites and could then give information about glucose metabolism and composition of lipids in adipose tissue. Use of other stable isotopes such as ^{2}H could be considered (210).

The potential of TMR for clinical diagnosis of disease states, observation of alterations in tissue metabolism, study of tissue energetics and enzyme kinematics, and of assessment of therapy is clearly considerable. Indeed, it has been suggested (211) that ^{31}P and ^{13}C NMR could accomplish the goals to which positron emission tomography (PET) is also oriented, with the advantage of non-hazardous operation. The disadvantage of TMR is the low sensitivity which does not approach the detection limits possible with CT, PET and fluoroscopy, and the slow data acquisition time which is ~ 100 min for an adequate signal-to-noise ratio.

1.5.3 *Structural studies*

Both X-ray scattering and absorption in the edge region provide extra information which can be used to investigate molecular structure. When the photon energy is far from an edge, the elastic scattering can be described by the atomic form factor F_0. When the incident energy is close to an edge, the atomic form factor must be replaced by

$$F = F_0 + f' + if''$$

where both f' and f'' depend on the photoelectric cross-section. For light elements, f' and f'' are negligible at X-ray energies, but the presence of an anomalously scattering heavy atom can give rise to large changes in F at

X-ray energies near an edge (212). This effect is widely used for structure determination of protein crystals.

In absorption, EXAFS gives information on the number and type of atoms surrounding the primary absorbing atom and their inter-atomic distances (190). There is no requirement that the medium should be crystalline. The measurement is element specific and sensitive to concentrations of 1 part in 10^4 in a region of diameter 0.3 nm from the absorbing atom. Large molecules containing metal atoms, e.g. metalloproteins and metalloenzymes, are regarded as ideal candidates for study by this method (213). Of particular interest are the study of a calcium ion environment in bone mineral and related calcium phosphates (214), an examination of changes in bone mineral structure during development (215) and a determination of the structure of bovine milk calcium phosphate (216).

EXAFS gives information about the radial distance of neighbouring atoms but does not give angular structural information. In contrast, XANES is sensitive to the correlation in position of many atoms and in principal can be used to determine bond angles and site symmetries, and hence can distinguish between different models of the local environment (195).

1.5.4 *Compton profiles and their relation to imaging measurements*

In recent years, inelastic (Compton) scattering of photons has been studied extensively (217). The observed Compton line for a bound electron is broadened by the Doppler term $\mathbf{p}\cdot\mathbf{q}$ where $\hbar p$ is the momentum of the bound electron and $\hbar q$ is the momentum transfer, and the wavelength shift is given by

$$\lambda_f - \lambda_i = (2h/mc)\sin^2\tfrac{1}{2}\theta + 2(\hbar p_z/mc)(\lambda_i\lambda_f)^{1/2}\sin\tfrac{1}{2}\theta$$

where p_z is the component of $\mathbf{p}$ along $\mathbf{q}$, which is taken to define the z-axis. The cross-section for Compton scattering contains the Compton profile (216, 217)

$$J(p_z) = \int_{-\infty}^{\infty}\int_{-\infty}^{\infty} dp_x dp_y \rho(p_x p_y p_z) \tag{1.43}$$

which is a one-dimensional projection of the electron -momentum distribution $\rho(\mathbf{P})$ and satisfies the sum rule

$$\int_{-\infty}^{\infty} dp_z J(p_z) = Z. \tag{1.44}$$

Compton profile measurements can be made on solid, liquid or gaseous samples and give information on a range of physical and chemical properties, including the influence of chemical bonding on the electron momentum distribution (219, 220). The most common procedure for comparison of experimental data with theoretical predictions is to take a predicted three-dimensional distribution and integrate over two momentum components to give a predicted profile (221). There is a direct analogy between the Fourier reconstruction method in tomography and the Fourier reconstruction method for momentum distributions. In tomography, the projection theorem states that the Fourier transform of the projection of the two-dimensional distribution of interest at a given orientation gives the Fourier transform of the distribution along the corresponding line through the origin. Similarly, the Fourier transform of the momentum distribution is

$$\rho(xyz) = (2\pi)^{-3} \iiint dp_x dp_y dp_z e^{i(p_x x + p_y y + p_z z)} \rho(p_x p_y p_z) \tag{1.45}$$

which, evaluated at $(0, 0, z)$ gives

$$\rho(0, 0z) = (2\pi)^{-1} \int_{-\infty}^{\infty} dp_z e^{ip_z z} J(p_z) \tag{1.46}$$

using eq. (1.43). Thus the Fourier transform of the measured projection gives the Fourier transform of the momentum distribution on a line along the z-axis. By varying the direction of the scattering vector with respect to the specimen the transform along several such lines is determined.

When positron annihilation measurements are made with a slit which is long in one direction, say the y-direction, and coincidences are detected as a function of the angle $\theta = \hbar p_z/mc$, the quantity determined is given by (222)

$$N(p_z) = \int_{-\infty}^{\infty} \int_{-\infty}^{\infty} dp_x dp_y \rho_+(\mathbf{p}) \tag{1.47}$$

where the integration over dp_x arises if an energy measurement is not made. Here ρ_+ is the momentum distribution of the electron-positron pair, so that the quantity $N(p_z)$ relates to $J(p_z)$ only if the positron wavefunction can be replaced by a constant. When a small slit or large position sensitive detectors are used, the quantity determined is

$$N(p_y p_z) = \int_{-\infty}^{\infty} dp_x \rho_+(\mathbf{p}) \tag{1.48}$$

In positron tomography, the deflections through angles θ, which are of the order of a few mrad, are neglected in the reconstruction process but are included in estimates of the resolution (223).

1.5.5 *Imaging methods in biopsy and pathology*

Oldendorf (224) has suggested that a fan beam of soft X-rays in the energy range 3–8 keV could be used to carry out computed tomography on a microscopic scale, utilizing the large photoelectric cross-section at these energies and taking dual energy measurements above and below K-edges for $Z = 19$–27. The application would be in biopsy and pathology, and would allow tissue specimens to be examined in a form close to the living state. For the reasons discussed in section 1.5.1, it is quite likely that measurements in the soft X-ray region are strongly influenced by molecular effects, which could modify interpretation of images and make comparisons difficult, especially in biological materials.

Acknowledgements. We are indebted to Mr. J. Grant for valuable comments and to Mrs. G. Wallace for typing the paper.

References

1. W. B. Gilboy and J. Foster, *Research Techniques in Non-Destructive Testing* (ed. R. S. Sharpe), Vol. 6, (1982) 255.
2. F. H. DeLand, E. E, Kim, F. J. Primus, M. E. Dine and D. M. Goldenberg, *AJR* **138** (1982) 145.
3. Radioimmunodetection of cancer workshop, *Cancer Res.* **40** (1980) 2957–3086.
4. S. Twomey, *J Franklin Inst.* **279** (1965) 95.
5. D. L. Phillips, *J. Assoc. Comp. Mach.* **9** (1962) 84.
6. T. S. Huang (ed.), *Picture Processing and Digital Filtering*, Applied Physics, Vol. 6, Springer-Verlag (1975).
7. A. Rosenfield, (ed.), *Digital Picture Analysis*, Applied Physics, Vol. 11, Springer-Verlag (1976).
8. G. T. Herman, *Image Reconstruction from Projections: the Fundamentals of Computerized Tomography*, Academic Press (1980).
9. L. A. Shepp, *J. Comput. Assist. Tomogr.* **4** (1980) 94.
10. A. Klug and R. A. Crowther, *Nature* **238** (1972) 435.
11. D. F. Jackson, *Proc. Int. Conf. on Applications of Physics to Medicine and Biology*, World Scientific Publishing (1982).
12. D. F. Jackson and D. J. Hawkes, *Physics Reports* **70** (1981) 169.
13. J. H. Hubbell, H. A. Gimm and I. Overbo, *J. Phys. Chem. Ref. Data* **9** (1980) 1023; S. Murugesu and D. F. Jackson, *Phys. Med. Biol.*, **27** (1983) 753.
14. G. T. Herman, *Phys. Med. Biol.* **24** (1979) 81.
15. R. A. Koeppe, R. M. Brugger, G. A. Schlapper, G. N. Larsen and R. J. Jost, *J. Comput. Assist. Tomogr.* **5** (1981) 79.
16. R. P. Kruger, G. W. Wecksung and R. A. Morris, *Opt. Eng.* **19** (1980) 273.
17. A. M. Cormack, *J. Appl. Phys.* **34** (1963) 2722.
18. H. Schomberg, *J. Phys. D. Appl. Phys.* **11** (1978) L181; G. H. Glover and J. C. Sharp, *IEEE Trans. Sonics and Ultrasonics* **SU-24** (1977) 229.
19. R. Maini, M. F. Iskander and C. H. Durney, *Proc. IEEE* **68** (1980) 1550.
20. H. Schomberg, in *Mathematical Aspects of Computerized Tomography*, G. T. Herman and F. Natterer (eds.), Springer-Verlag (1981).

21. P. D. Edmonds (ed.), *Ultrasonics*, Academic Press (1981).
22. G. Wade, R. K. Mueller and M. Kaveh, in *Computer Aided Tomography and Ultrasonics in Medicine*, J. Raviv, J. F. Greenleaf and G. T. Herman (eds.), North Holland (1979).
23. G. C. McKinnon and R. H. T. Bates, *Ultrasonic Imaging* **2** (1980) 48.
24. R. H. T. Bates, G. C. McKinnon and A. D. Seagar, *IEEE Trans. Biomed. Eng.* **BME-27** (1980) 418.
25. R. Henderson and J. G. Webster, *IEEE Biomed. Eng.* **BME-25** (1978) 250.
26. L. R. Price, *IEEE Trans. Nucl. Sci.* **NS-26** (1979) 2736.
27. P. S. Rao, K. Santosh and E. C. Gregg, *Radiology*, **135** (1980) 769.
28. K. A. Dines and R. J. Lytle, *Proc. IEEE* **67** (1979) 1065.
29. F. H. DeLand, *J. Nucl. Med.* **23** (1982) 73.
30. R. F. Brown, *IEEE Trans Biomed. Eng.* **BME-27** (1980) 1.
31. H. H. Barrett and F. A. Horrigan, *Appl. Opt.* **12** (1973) 2686.
32. R. S. May, Z. Akcasu and G. F. Knoll, *Appl. Opt.* **13** (1974) 2589.
33. T. F. Budinger, *J. Nucl. Med.* **21** (1980) 579.
34. R. J. Jaszczak, R. E. Coleman and C. B. Lim, *IEEE Trans. Nucl. Sci.* **NS-27** (1980) 1137.
35. M. M. Ter-Pogossian, N. A. Mullani, D. C. Ficke, J. Markham and D. L. Snyder, *J. Comput. Assist. Tomogr.* **5** (1981) 227.
36. Proc. Workshop on "Time-of-flight tomography", 1982, to be published by IEEE Computer Society.
37. T. F. Budinger, G. T. Gullberg and R. H. Huesman, in *Image Reconstruction from Projections—Implementations and Applications*, G. T. Herman (ed.), Springer-Verlag (1979) 147.
38. G. T. Gullberg and T. F. Budinger, *IEEE Trans. Biomed. Eng.* **BME-28** (1981) 142.
40. C. H. Jones, *Phys. Med. Biol.* **27** (1982) 463.
41. A. H. Barrett, P. C. Myers and N. I. Sadowsky, *Am. J. Roentgen.* **134** (1980) 365.
42. T. C. Farrar and E. D. Becker, *Pulse and Fourier Transform NMR, Academic Press* (1971).
43. *Proc. Royal Society meeting: Phil. Trans. Roy. Soc. Lond.* **289** (1980) 379; R. L. Witcofski, N. Karstaedt and C. L. Partain (eds.), *Proc. Int. Symp. on Nuclear Magnetic Resonance Imaging*, Bowman Gray School of Medicine of Wake Forest University (1982).
44. L. Kaufman, L. E. Crooks and A. R. Margulis (eds.). *Nuclear Magnetic Resonance Imaging in Medicine*, Igaku-Shoin (1982).
45. C. F. Bore, Ph.D Thesis, University of Surrey (1982).
46. I. R. Young, D.R. Bailes, A. G. Collins and D. J. Gilderdale, *Proc. Int. Symp. on NMR Imaging*, Bowman Gray School of Medicine of Wake Forest University (1982) 93.
47. L. Crooks, M. Arakawa, J. Hoenninger, J. Watts, R. McRee, L. Kaufman, P. L. Davis, A. R. Margulis and J. DeGroot, *Radiology* **143** (1983) 169.
48. L. Crooks, P. Sheldon, L. Kaufman, W. Rowan and T. Miller, *IEEE Trans. Nucl. Sci.* **NS-29** (1982) 1181.
49. J. C. Gore, *Proc. Int. Symp. on NMR Imaging*, Bowman Gray School of Medicine of Wake Forest University (1982) 15.
50. D. G. Taylor and C. Bore, *J. Computer Tomography* **5** (1981) 122.
51. L. Kaufman, *Proc. Int. Conf. in Applications of Physics to Medicine and Biology*, World Scientific Publishing (1982).
52. M. G. Silk, *Research Techniques in Non Destructive Testing* (ed. R. S. Sharpe), Vol 3 (1977) 51.
53. P. T. Wells, *Br. Med. Bull.* **36** (1980) 257.
54. T. F. Budinger, *Proc. Int. Symp. on NMR Imaging*, Bowman Gray School of Medicine of Wake Forest University (1982) 51.
55. K. J. W. Taylor, *An Atlas of Ultrasonography* (1978).
56. J. C. Rodger, *Br. Med. Bull.* **36** (1980) 261.
57. T. P. Naidich, C. J. Moran, R. M. Pudlowski and J. Hannaway, *Medical Clinics of N. America* **63** (1979) 849.
58. F. F. Hopkins, I. L. Morgan, H. D. Ellinger, R. V. Klinksiek, G. A. Meyer and J. N. Thompson, *IEEE Trans. Nucl. Sci.* **NS-25** (1981) 1717.
59. W. B. Gilboy, J. Foster and M. Folkard, *Nucl. Inst. & Meth.* **193** (1982) 209;

W. B. Gilboy, J. Foster, M. Folkard and A. A. Tajuddin, *Proc. 2nd Int. Symp. on Radiation Physics*, Penang (1982).
60. R. P. Kruger, G. W. Wechsung and R. A. Morris, *Optical Eng.* **19** (1981) 273.
61. F. G. Sommer, M. P. Capp, C. A. Tobias, E. V. Benton, K. H. Woodruff, R. P. Henke, W. Holley and H. K. Genant, *Invest. Radiol.* **13** (1978) 163.
62. J. Saudinos, *Proc. Int. Conf. in Applications of Physics to Medicine and Biology*, World Scientific Publishing (1982).
63. G. M. Bydder and R. E. Steiner, *Neuroradiology* **23** (1982) 231.
64. F. W. Smith, *Proc. Int. Symp. on NMR Imaging*, Bowman Gray School of Medicine of Wake Forest University (1982) 125.
65. G. M. Bydder, R. E. Steiner, I. R. Young, A. S. Hall, D. J. Thomas, J, Marshall, C. A. Pallis and N. J. Legg, *Am. J. Roentgen.* **139** (1982) 215.
66. M. P. Capp, *Radiology* **138** (1981) 541.
67. C. A. Mistretta, A. B. Crummy and C. M. Strother, *Radiology* **139** (1981) 273.
68. B. P. Drayer, E. R. Heinz, M. Dujoray, S. K. Wolfson and D. Gur, *J. Comput. Assist. Tomogr.* **3** (1979) 633; E. R. Heinz, P. Dubois, D. Osborne, B. Drayer and W. Barret, *J. Comput. Assist. Tomogr.* **3** (1979) 641; S. W. Young M. A. Noon, and M. Nassi, *J. Comput. Assist. Tomogr.* **4** (1980) 168; M. Heller, E. Grabbe and E. Bücheler, *Fortschr. Röntgenstr.* **134** (1981) 16.
69. B. H. Brundage, M. J. Lipton, R. J. Herfkens, W. H. Berninger, R. W. Redington, K. Chatterjee and E. Carlsson, *Circulation* **61** (1980) 826.
70. A. J. Hall, *Br. Med. Bull.* **36** (1980) 267.
71. J. P. Woodcock, *Br. Med. Bull.* **36** (1980) 243.
72. P. Atkinson and J. P. Woodcock, *Doppler Ultrasound and Its Clinical Uses* (1981).
73. A. K. Freimanis, in *Medical Imaging Techniques*, eds. K. Preston, K. J. W. Taylor, S. A. Johnson and W. R. Ayers (1979) 105.
74. C. R. Hill, *Proc. Int. Conf. in Applications of Physics to Medicine and Biology*, World Scientific Publishing Co. (1982).
75. V. R. McCready, *Br. Med. Bull.* **36** (1980) 209.
76. K. J. W. Taylor, D. Sullivan, J. Simeone and A. T. Rosenfeld, *Medical Imaging Techniques*, eds. K. Preston, K. J. W. Taylor, S. A. Johnson and W. R. Ayers (1979) 55.
77. H. H. Barrett, T. Bowen, R. S. Hershel, S. K. Gordon and D. A. Lelise, *Optical Society of America* (1975) (see ref. 134).
78. K. M. Hanson, Los Alamos Report LA-UR 78–1827 (1978), LA-7107-MS (1978); K. M. Hanson, J. N. Bradbury, T. M. Cannon, R. L. Hutson, O. B. Laubacher, R. J. Macek, M. A. Paciotti and C. A. Taylor, *Phys. Med. Biol.* **26** (1981) 25.
79. A. A. Mustafa and D. F. Jackson, *Phys. Med. Biol.* **26** (1981) 461.
80. V. W. Steward and A. M. Koehler, *Radiology* **110** (1974) 217.
81. J. Fabrikant, C. A. Tobias, M. P. Capp, E. V. Benton and W. R. Holley, *SPIE* **233** (1980) 255.
82. F. H. Doyle, J. C. Gore, J. M. Pennock, G. M. Bydder, S. J. Orr, R. E. Steiner, I. R. Young, M. Burl, H. Clows, D. J. Gilderdale, D. R. Bailes and P. E. Walters, *Lancet* **2** (1981) 53.
83. V. W. Steward and A. M. Koehler, *Nature* **245** (1973) 38; V. W. Steward and A. M. Koehler, *Surgical Neurology* **2** (1974) 823.
84. V. W. Steward, *J. Neurol. Sci.* **39** (1978) 261.
85. I. R. Young, A. S. Hall, C. A. Pallis, N. J. Legg, G. M. Bydder and R. E. Steiner, *Lancet* **2** (1981) 1063.
86. J. P. Neilson and V. D. Hood, *Br. Med. Bull.* **36** (1980) 249.
87. S. N. Rasmussen, *Medical Imaging Techniques*, eds. K. Preston, K. J. W. Taylor, S. A. Johnson and W. R. Ayers (1979) 175.
88. J. K. Udupa, *Computer Graphics and Image Processing* **17** (1981) 52.
89. S. B. Hemsfield, J. Fulenwider, B. Nordlinger, R. Barlow, P. Sones and M. Kutner, *Ann. Intl. Med.* **90** (1979) 185.
90. R. L. Walser and L. V. Ackerman, *J. Comput. Asst. Tomogr.* **1** (1977) 117.
91. J. E. Husband, K. J. Cassell, M. J. Peckham and J. S. Macdonald, *Br. J. Radiol.* **15** (1981) 50;

J. E. Husband, D. J. Hawkes and M. J. Peckham, *Radiology* **144** (1982) 553.
92. G. Williams, G. Bydder and L. Kreel, *Br. Med. Bull.* **36** (1980) 279.
93. A. Dutreix, *Advances in Radiation Protection and Dosimetry in Medicine*, eds. R. H. Thomas and V. Perez-Mendez (1980) 557.
94. H. K. Liu, *Computer Graphics and Image Processing* **6** (1977) 123; G. T. Herman and G. T. Liu, *Computer Graphics and Image Processing* **9** (1979) 1; G. T. Herman, J. K. Udupa, D. M. Kramer, P. C. Lauterbur, A. M. Rudin and J. S. Schneider, *SPIE* **273** (1981) 35.
95. J. R. Phillips, B. K. Barnes, M. L. Barnes, D. K. Hamlin and E. G. Medina-Ortega, *EPRI Report* **NP-1952** (1981).
96. D. E. Kuhl, J. Engel, M. E. Phelps and C. Selin, *Ann. Neurol.* **8** (1980) 348.
97. M. E. Phelps, *Sem. Nucl. Med.* **11** (1981) 32.
98. K. F. Hübner, J. T. Purvis, S. M. Mahaley, J. T. Robertson, S. Rogers, W. D. Gibbs, P. King and C. L. Partain, *J. Comput. Assist. Tomogr.* **6** (1982) 544.
99. W. H. Oldendorf, *J. Nucl. Med.* **19** (1978) 1182.
100. G. S. Keyes, N. J. Pelc, S. J. Riederer, L. E. Sieb and D. R. Enzmann, *Digital Fluorography : a Technology Update*, GE publication 5312 (1981); W. R. Brody (ed.) *Proc. SPIE* **314** (1981).
101. J. H. Bürsch, H. J. Hahne, R. Brennecke, D. Grönemeier and P. H. Heintzen, *Radiol.* **141** (1981) 39.
102. J. H. Battocletti, R. E. Halbach, S. X. Salles-Cunha and A. Sances, *Med. Phys.* **8** (1981) 435.
103. T. J. Pedley, *The Fluid Mechanics of Large Blood Vessels*, Cambridge University Press (1958).
104. S. Antman, in *Dye Curves : the Theory and Practice of Indicator Dilution*, D. A. Bloomfield (ed.), University Park Press (1974).
105. L. Axel, *Radiology.* **137** (1980) 679.
106. M. J. Lipton, B. Brundage, P. W. Doherty, R. Herfkens, W. H. Berninger, R. W. Redington, K. Chatterjee and E. Carlsson, *Cardiovascular Med.* **4** (1979) 1219.
107. S. K. Hilal, in *Radiology of the Skull and Brain: Angiography*, T. H. Newton and D. G. Potts (eds.), C. V. Mosby Co. (1974).
108. K. E. Britton, *Br. Med. Bull.* **36** (1980) 215.
109. S. X. Salles-Cunha, R. E. Halbach, J. H. Battocletti and A. Sances, *Med. Phys.* **9** (1982) 188.
110. J. R. Singer in *Nuclear Magnetic Resonance Imaging*, Saunders (1982), Chapter 12.
111. R. A. Robb, E. L. Ritman, B. K. Gilbert, J. H. Kinsey, L. D. Harris and E. H. Wood, *IEEE Trans. Nucl. Sci.* **NS-26** (1979) 2713.
112. D. P. Boyd, R. G. Gould, J. R. Quinn, R. Sparks, J. H. Stanley and W. B. Hermannsfeldt, *IEEE Trans. Nucl. Sci.* **NS-26** (1979) 2724.
113. L. Crooks, M. Arakawa, J. Hoenninger, J. Watts, R. McRee, L. Kaufman, P. L. Davis, A. R. Margulis and J. DeGroot, *Radiology* **143** (1982) 169.
114. R. A. Kruger, *Proc Int. Conf. in Applications of Physics to Medicine and Biology*, World Scientific Publishing Co. (1982).
115. A. C. Devuono, P. A. Schlosser, F. A. Kulacki and P. Munshi, *IEEE Trans. Nucl. Sci.* **NS-27** (1980) 814.
116. P. Konsky, *New Scientist* 9 April (1981) 8.
117. D. G. Boyd, in *Radiology of the Skull and Brain: Technical Aspects of Computer Tomography.* T. H. Neuton and D. G. Potts (eds.), C. V. Mosby Co. (1981).
118. W. H. Berninger, R. W. Redington, P. Doherty, M. J. Lipton and E. Carlsson, *J. Comput. Assist. Tomogr.* **3** (1979) 155.
119. S. L. Bacharach and M. V. Green, *IEEE Trans. Nucl. Sci.* **NS-29** (1982) 1343.
120. H. G. Ostrow, S. Allen, S. L. Bacharach, *J. Nucl. Med.* **20** (1979) 610.
121. P. D. Esser (ed.), *Functional Mapping of Organ Systems*, Society of Nuclear Medicine (1981).
122. F. H. Doyle, J. M. Pennock, L. M. Banis, M. J. McDonnell, G. M. Bydder, R. E. Steiner, I. R. Young, C. J. Clarke, T. Pasmore and D. J. Gilderdale, *Am. J. Roentgen.* **138** (1982) 193.
123. F. W. Smith, J. R.Mallard, A. Reid and J. M. S. Hutchison, *Lancet* (1981) 963.
124. R. A. Brooks, G. DiChiro and M. R. Keller, *J. Comput. Assist. Tomogr.* **4** (1980) 489.
125. A. M. Koehler, J. G. Dickinson and W. M. Preston, *Rad. Res.* **26** (1965) 334.

126. C. A. Tobias, J. I. Fabrikant, E. V. Benton and W. R. Holley, *Report LBL*-11220 (1980) 335.
127. A. A. M. Mustafa and D. F. Jackson, *Phys. Med. Biol.* **28** (1983) 169.
128. ICRP, *Report on Task Group on Reference Man* (1959); ICRU, *Report* 10*b* (1964), *Report* 16 (1970); *Science Data Book*, ed. R. M. Tennant (1971); C. Constantinou, Ph.D Thesis, University of London (1978); D. R. White, *Med. Phys.* **5** (1978) 467.
129. A. M. Koehler and H. Berger, *Research Techniques in Non Destructive Testing*, ed. R. S. Sharpe, **2** (1973) 1.
130. V. W. Steward, *IEEE Trans. Nucl. Sci.* **NS-23** (1976) 577.
131. J. W. Motz and M. Danos, *Med. Phys.* **5** (1978) 8.
132. A. A. M. Mustafa, Ph.D Thesis, University of Surrey (1981).
133. S. L. Kramer, D. R. Moffet, R. L. Martin, E. P. Colton and V. W. Steward, *Radiology* **135** (1980) 485.
134. H. M. Barrett, T. Bowen, R. S. Hershel, S. K. Gordon and D. A. Delise, *Tech. Digest on Image Processing for* 2-*D and* 3-*D Reconstruction from Projections*, Optical Soc. of America **MAl-1** (1975).
135. W. T. Scott, *Rev. Mod. Phys.* **35** (1963) 231.
136. D. West and A. C. Sherwood, *Nature* **239** (1972) 157; Report AERE-R 7190 (1972).
137. D. West and A. C. Sherwood, *Non Destructive Testing* **6** (1973) 249.
138. J. Saudinos, G. Charpak, F. Sauli, D. Townsend and J. Viniciarelli, *Phys. Med. Biol.* **20** (1975) 890.
139. K. Kouris, N. M. Spyrou and D. F. Jackson, *Research Techniques in Non Destructive Testing*, ed. R. S. Sharpe, Vol. 6, 211 (1982).
140. J. J. Battista and M. J. Bronskill, *Phys. Med. Biol.* **26** (1981) 81.
141. A. Engström, *Physical Techniques in Biological Research*, ed. A. W. Pollister, *III A* (1966) 87; D. K. Bowen, *Ann. N. Y. Acad. Sci.* **342** (1980) 22; J. Kirz and D. Sayre, *Synchrotron Radiation Research*, ed. H. Winick and S. Doniach (1980) 277; G. Schmahl, D. Rudolph and B. Niemann, *Scanned Image Microscopy*, ed. E. A. Ash (1980) 393; E. Spiller, *Scanned Image Microscopy*, ed. E. A. Ash (1980) 365.
142. I. V. Mitchell and K. M. Barfoot, *Nucl. Sci. Appl.* **1** (1981) 99.
143. F. Bosch, A. E. L. Goresy, W. Herth, B. Martin, R. Nobling, B. Povh, H. D. Reiss and K. Traxel, *Nucl. Sci. Appl.* **1** (1980) 33.
144. R. E. Snyder and D. C. Secord, *Phys. Med. Biol.* **27** (1982) 515.
145. T. Grönberg, T. Almén, K. Golman, K. Lidén, S. Mattsson and S. Sjöberg, *Phys. Med. Biol.* **26** (1981) 501.
146. L. Ahlgren and S. Mattsson, *Phys. Med. Biol.* **26** (1981) 19.
147. P. Bloch, G. Garavaglia, G. Mitchell and I. M. Shapiro, *Phys. Med. Biol.* **20** (1976) 56; L. Ahlgren and S. Mattsson, *Phys. Med. Biol.* **24** (1979) 136.
148. K. Kouris and N. M. Spyrou, *Nucl. Inst. and Meth.* **153** (1978) 477.
149. K. Kouris, N. M. Spyrou and D. F. Jackson, *Nucl. Inst. and Meth.* **187** (1981) 539.
150. P. Puumalainen, A. Vimarihuhta, H. Olkkonen and E. M. Alhava, *Phys. Med. Biol.* **27** (1982) 425; H. Olkkonen, P. Puumalainen, P. Karjalainen, and E. M. Alhava, *Invest. Radiol.* (1981).
151. S. A. Kerr, K. Kouris, C. E. Webber and T. J. Kennett, *Phys. Med. Biol.* **25** (1980) 1037.
152. I. Isherwood, R. A. Rutherford, B. R. Pullan and P. H. Adams, *Lancet* **2** (1976) 712; H. K. Genant and D. Boyd, *Invest. Radiol.* **12** (1977) 545; C. S. Revak, *J. Comput. Assist. Tomogr.* **4** (1980) 342.
153. V. Elsasser and J. Reeve, *Br. Med. Bull.* **36** (1980) 293.
154. P. Ruegsegger, V. Elsasser, M. Anliker, H. Gnehm, H. Kind and A. Prader, *Radiology* **121** (1976) 93.
155. G. Deconninck, *Introduction to Radioanalytical Physics*, Elsevier (1978).
156. M. Hyvönen-Dabek, M. Riihonen and J. T. Dabek, *Phys. Med. Biol.* **24** (1979) 988.
157. L. Zikovsky and E. A. Schweikert, *J. Radioanal. Chem.* **37** (1977) 571.
158. E. A. Schweikert, *Some Selected Nuclear Techniques in Research and Development*, IAEA (1978) 1; N. M. Spyrou and I. P. Matthews, *Int. J. App. Rad. Isot.* **33** (1982) 61.
159. J. R. McGinley, G. J. Stock, E. A. Schweikert, J. B. Cross, R. Zeisler and L. Zikovsky, *J. Radioanal. Chem.* **43** (1978) 559.
160. S. H. Cohn, *Med. Phys.* **8** (1981) 145.

161. N. M. Spyrou, *J. Radioanal. Chem.* **61** (1981) 211; N. M. Spyrou, G. E. Nicolaou, I. P. Matthews, L. G. Stephens-Newsham and I. Othman, *J. Radioanal. Chem.* **71** (1982) 519.
162. R. F. Eilbert, A. M. Koehler, J. M. Sisterson, R. Wilson and S. J. Adelstein, *Phys. Med. Biol.* **22** (1977) 817.
163. D. Brune, V. Lindh and H. Lundqvist, *Analytica Chimica Acta* **89** (1977) 267.
164. K. Kyere, B. Oldroyd, C. B. Oxby, L. Burkinshaw, R. E. Ellis and G. L. Hill, *Phys. Med. Biol.* **27** (1982) 805.
165. D. Vartsky, B. J. Thomas, D. J. Hawkes and J. H. Fremlin, *Phys. Med. Biol.* **21** (1976) 970.
166. D. Vartsky, K. J. Ellis, D. M. Hull and S. H. Cohn, *Phys. Med. Biol.* **24** (1979) 689.
167. H. Berger, *Research Techniques in Non Destructive Testing*, ed. R. S. Sharpe, **1** (1970) 269.
168. H. L. Atkins, *Materials Evaluation* **23** (1965) 453.
169. R. P. Kruger and R. A. Morris, *SPIE* **182** (1979) 158; N. M. Spyrou, private communication.
170. R. E. Alvarez and A. Macovski, *Phys. Med. Biol.* **21** (1976) 733.
171. R. A. Brooks, *J. Comput. Assist. Tomogr.* **1** (1977) 487.
172. R. T. Ritchings and B. R. Pullan, *J. Comput. Assist. Tomogr.* **3** (1979) 842.
173. R. A. Rutherford, B. R. Pullan and I. Isherwood, *Neuroradiology* **11** (1976) 15.
174. G. Williams, G. M. Bydder and L. Kreel, *Br. Med. Bull.* **36** (1980) 279.
175. H. W. Kraner, J. L. Alberi and H. L. Atkins, *IEEE Trans. Nucl. Sci.* **NS-20** (1972) 389.
176. J. L. Alberi, H. W. Kraner, P. Bradley-Moore and H. L. Atkins, *IEEE Trans. Nucl. Sci.* **NS-21** (1974) 635.
177. S. J. Riederer and C. A. Mistretta, *Med. Phys.* **4** (1977) 474.
178. R. A. Kruger, J. D. Armstrong, J. A. Sorenson and L. T. Niklason, *Radiology* **140** (1981) 213.
179. S. J. Riederer, R. A. Kruger and C. A. Mistretta, *Med. Phys.* **8** (1981) 54.
180. S. J. Riederer, R. A. Kruger and C. A. Mistretta, *Med. Phys.* **8** (1981) 471.
181. S. J. Riederer, R. A. Kruger, C. A. Mistretta, D. L. Ergun and C. G. Shaw, *Med. Phys.* **8** (1981) 480.
182. R. Birch, M. Marshall and G. M. Ardran, *Catalogue of Spectral Data for Diagnostic X-rays*, Hospital Physicists' Association (1979).
183. D. F. Jackson and D. J. Hawkes, *Phys. Med. Biol.* **28** (1983) 289.
184. R. W. G. Chapman, G. Williams, G. Bydder, R. Dick, S. Sherlock and L. Kreel, *Br. Med J.* **280** (1980) 440.
185. R. A. Brooks and G. D. Chiro, *Radiology* **126** (1978) 255.
186. H. L. Atkins, W. Hauser and H. W. Kraner, *Am. J. Roentgen.* **114** (1972) 176.
187. J. Kirz, *Ann. N. Y. Acad. Sci.* **342** (1980) 273.
188. J. Williamson and D. E. Watt, *Phys. Med. Biol.* **17** (1972) 486; D. I. Thwaites and D. E. Watt, *3rd Symp. on Neutron Dosimetry in Medicine and Biology*, Euratom 5848 (1977) 45.
189. R. D. Deslattes, *Acta Cryst.* **A25** (1969) 89.
190. C. D. Garner, *Proc. Daresbury Study Weekend* DL/SCI/R13 (1979) 7.
191. P. M. Atha, P. C. Ford, C. D. Garner, A. A. MacDowell, I. M. Hillier, M. F. Guest and V. R. Saunders, *Chem. Phys. Lett.* (1981).
192. J. M. Hubbell, *J. Physique* **32** (1971) C4.
193. J. Stöhr, R. Jaeger, J. Feldhaus, S. Brennan, D. Norman and G. Apai, *Appl. Opt.* **19** (1980) 3911.
194. A. Kotani and Y. Toyozawa, *Synchrotron Radiation*, ed. C. Kunz, Springer Verlag, Berlin (1979) Chapter 4.
195. G. R. Greaves, P. J. Durham, G. Diakun and P. Quinn, *Nature* **294** (1981) 139; P. J. Durham, J. B. Pendry and C. A. Hodges, *Comput. Phys Commun.* (1981).
196. D. J. Hawkes and D. F. Jackson, *Phys. Med. Biol.* **25** (1980) 1167.
197. R. H. Pratt, A. Ron and H. K. Tseng, *Rev. Mod. Phys.* **45** (1973) 273.
198. Sung Dahm Oh, J. McEnnan and R. H. Pratt, *Phys. Rev.* **A14** (1976) 1428.
199. J. H. Hubbell, Wm. J. Veigele, E. A. Briggs, R. T. Brown, D. T. Cromer and R. J. Howerton, *J. Phys. Chem. Ref. Data* **4** (1975) 471.
200. L. Kissel and R. H. Pratt, *Phys. Rev. Lett.* **40** (1978) 387; L. Kissel, R. H. Pratt and S. C. Roy, *Phys. Rev.* **A22** (1980) 1970.
201. F. Folkman, *NATO Advanced Study Institute Series B, Physics* **28** (1978) 239.
202. G. Bigginger, J. M. Joyce, J. A. Tanis and S. L. Varghose, *Phys. Rev. Lett.* **44** (1980) 241.

203. J. Tousset, *NATO Advanced Study Institute Series B, Physics* **28** (1978) 101.
204. Oxford Research Systems, *TMR Spectroscopy—Product and Application, Note* (1980).
205. R. E. Gordon, P. E. Hanley and D. Shaw, *Progress in NMR Spectroscopy* **15** (1982) 1.
206. G. K. Radda, L. Chan, P. B. Bore, D. G. Gadian, B. D. Ross, P. Styles and D. Taylor, *Proc. Int. Symp. on NMR Imaging*, Bowman Gray School of Medicine of Wake Forest University (1982) 159.
207. B. D. Ross, G. K. Radda, D. G. Gadian, G. Rocker, M. Esiri and J. Falconer-Smith, *New Eng. J. Med.* **304** (1981) 1338.
208. R. H. T. Edwards, M. J. Dawson, D. R. Wilkie, R. E. Gordon and D. Shaw, *Lancet*, March 27 (1982) 725.
209. T. J. Brady, C. T. Burt, M. R. Goldman, I. C. Pykett, F. S. Buonanno, J. P. Kistler, J. H. Newhouse, W. S. Hinshaw and G. M. Pohost, *Proc. Int. Symp. on NMR Imaging*, Bowman Gray School of Medicine of Wake Forest University (1982) 175.
210. D. F. Jackson, *Proc. Int. Conf. in Applications of Physics to Medicine and Biology*, World Scientific Publishing (1982).
211. R. L. Nunnally, *Proc. Int. Symp. on NMR Imaging*, Bowman Gray School of Medicine of Wake Forest University (1982) 181.
212. J. Helliwell, *Proc. Daresbury Study Weekend* DL/SC1/R13 (1979) 1.
213. I. H. Munro, *Chemistry in Britain* **15** (1979) 330; S. S. Hasnain, N. J. Blackburn, J. Bordas, G. P. Diakun, C. D. Garner, P. F. Knowles, M. H. J. Koch, R. M. Miller and J. C. Phillips, *J. Am. Chem. Soc.* (1981); C. D. Garner, S. S. Hasnain, I. Bremner and J. Bordas, Daresbury Laboratory Preprint (1981).
214. R. M. Miller, D. W. L. Hukins, S. S. Hasnain and P. Lagarde, *Biochem. Biophys. Res. Commun.* **99** (1981) 102.
215. N. Binstead, S. S. Hasnain and D. W. L. Hukins, *Biochem. Biophys. Res. Commun.* (1982).
216. C. Holt, S. S. Hasnain and D. W. L. Hukins, *Nature* (1981); *Biochim. Biophys. Acta* (in the press).
217. M. J. Cooper, *Contemp. Phys.* **18**, (1977) 489.
218. P. Platzman and N. Tzoar, *Compton Scattering*, ed. B Williams, McGraw-Hill (1977) Chapter 2.
219. I. R. Epstein and A. C. Tanner, *Compton Scattering*, ed. B Williams, McGraw-Hill (1977) Chapter 7.
220. R. S. Holt and M. J. Cooper, *Mol. Phys.* **40** (1980) 857.
221. P. E. Mijnarends, *Compton Scattering*, ed. B. Williams, McGraw-Hill (1977) Chapter 10.
222. S. Berko, *Compton Scattering*, ed. B. Williams, McGraw-Hill (1977) Chapter 9.
223. A. P. Jeavons, D. W. Townsend, N. L. Ford, K. Kull, A. Manvel, O. Fischer and M. Peter, *IEEE Trans Nucl. Sci.* **NS-25** (1978) 164.
224. W. H. Oldendorf, *J. Comp. Assist. Tomogr.* **4** (1980) 141.

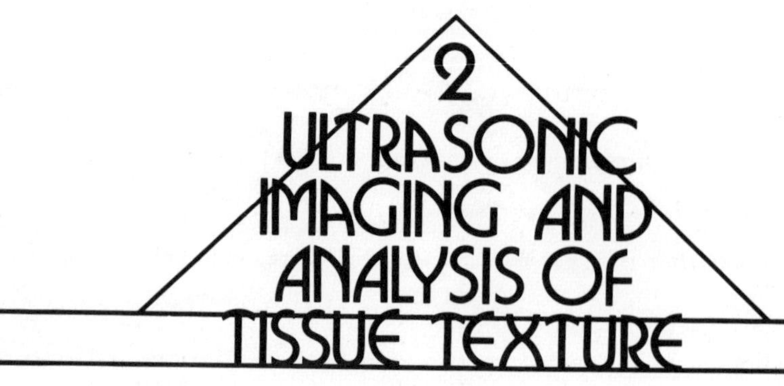

2 ULTRASONIC IMAGING AND ANALYSIS OF TISSUE TEXTURE

R.C. Chivers

2.1 Introduction

The vast literature on ultrasonic techniques in diagnostic medicine (1) is suggestive of the increasingly widespread use and value of the techniques and also of the need for their further development. The overwhelming majority of the techniques are based on pulse echo imaging. It is the purpose of this contribution to draw together briefly the main threads of our current understanding of the scientific basis on which the imaging process is based, and thus to identify some of the areas in which future work might be most effective in the search for optimal extraction and display of the information available. Before the more detailed discussion is presented, it is perhaps important to define its conceptual context.

2.1.1 *Developments in medical and ultrasonic imaging*

The very early ill-controlled therapeutic applications of ultrasound in medicine were largely discontinued in 1939 (2) which made their revival conditional on a greater understanding of the mechanisms of interaction of the ultrasound and the tissue. Subsequent failure to parallel the diagnostic process of taking transmission X-ray pictures of the head by using transmission ultrasonography in 1947 (3) epitomizes the weakness of the empirical approach. Fortunately the pulse-echo technique of visualization developed by Wild and Reid in 1952 (4) proved more fortunate in its choice of site and provided the basis for the majority of commercial instruments in use today. Many alternative methods of ultrasonic imaging have been suggested (5, 6) but none have yet achieved comparable status in the medical field as this early technique and its technological heirs.

The initial value of pulse-echo imaging lay in the tomographic visualization it provided, and the ability to visualize soft tissue structures without the use of contrast medium or ionizing radiation. The story of subsequent developments is not one of uniform progress. The interpretation of the original success of the pulse-echo process had two important features. Firstly, the production of recognizable anatomical cross-sections implied that the velocity of ultrasound in soft tissues was approximately constant. Secondly, the model invoked for the origin of the echoes from organ boundaries was the simplest available. It was that of a plane wave impinging normally on an infinitely extended, infinitely thin interface between two infinitely extended homogeneous lossless media—see, for example, (7). The evidence for the first of these which was available prior to the experimental implementation of the technique is summarized in table 2.1 (from (8)). It can be seen that it was inadequate by almost any standards, although subsequent laboratory work has given a clearer but still indistinct indication of the validity of the assumption (8, 9). The justification for the thin interface model remains sparse, even today. The obviously invalid assumptions of the model, specifically the experimental difficulty of obtaining plane waves, and the inhomogeneous lossy nature of most soft tissues (8, 9), have not apparently stimulated experimental investigations of a controlled nature. Ahuja (12) has produced some theoretical computations based on a Voigt model for the tissues (13). While this has demanded the inclusion of the attenuation coefficients for the tissues (see section 2.2.2), in order to determine the constants of the model, the author gives no report of associated experiments. Thus the main developments of the imaging process were primarily technological (14). The display of these relatively strong echoes that were believed to originate from the organ boundaries was optimized both by compounding the scan (15), and by signal processing which enhanced their display to the exclusion of other 'noise' (e.g. (16)).

The conceptual three-dimensional visualization offered by ultrasonic holography (17) was unfortunately limited by fundamental considerations that emerged with its implementation. Nevertheless it demonstrated very clearly the value of real time imaging and thus catalysed the technological innovations needed to provide real-time pulse-echo scanners (18, 19). As X-ray computerized tomography infringed the erstwhile monopoly of ultrasound for producing anatomical cross-sections of soft tissue regions, so ultrasound moved ahead into real-time visualization, with the additional advantages of

Table 2.1 Data on the velocity of ultrasound in soft tissues prior to 1952 (8).

Date	*Tissue*	*Velocity* (m/s)	*Reference*
1949	human biceps muscle	1515–1587	Ludwig and Struthers (10).
	human calf muscle	1500–1610	"
	human quadriceps muscle	1504–1563	"
1950	dog brain	1515	Ludwig (11).
	dog kidney	1558	"
	human limb	1540	"

lower cost, quicker examination, and increasing assurance of minimal potential hazard.

In spite of the efforts mentioned above devoted to enhancing the display of tissue boundaries at the expense of any other information in the pictures, there were continuing reports in the literature that the texture of the interior of the tissue volumes appeared to be of diagnostic value. The atlas produced by Howry in 1965 (20) remains impressive. Other evidence of this type has been summarized elsewhere (21). The cumulation of this evidence led eventually to two developments of importance. The first was the introduction of a new philosophy in signal processing in which the large dynamic range in the signals in the echo train was compressed so that even the smallest amplitudes could be displayed on a grey-scale scanner (22) alongside the reflections from the anatomical boundaries. The second was the growth of attempts to quantify these textural echoes, and led to the emergence of the whole field of tissue characterization (23).

Whereas the strong boundary echoes corresponded to short wavelength scattering (the ray approximation), it was believed that the textural echoes arose from smaller scale structures (i.e. diffractive scattering or Rayleigh scattering)(24). In the field of industrial non-destructive testing, where a similar physical problem arises in connection with the scattering of ultrasound from granular materials (21), the three scattering regimes are sometimes referred to as 'diffusion scattering'; 'stochastic scattering' and 'Rayleigh scattering' respectively (25). The identity of the scattering components of soft tissues that give rise to the textured features of the images produced has been the subject of much speculation over the last ten years, and appears at present still to be ill-defined. It is hoped that the discussion following will at least help to clarify some of the questions that remain to be answered in linking the acoustic structure of the tissue to the texture displayed. The discussion will be restricted to imaging by pulse-echo techniques.

2.1.2 *The components of a pulse-echo imaging system*

The pathway from the scattering interaction, that may be supposed to give rise to the displayed texture, to the texture itself is fraught with potential variation. A schematic diagram is shown in figure 2.1, in which the features labelled may each contribute to the texture of the image finally displayed. The parameters of variation may be broadly classified into three groups: acoustic, signal processing and display.

The acoustic section is worthy of particular attention since it incorporates most of the appropriate physics and also those aspects over which there is relatively little control. In addition to the scattering process which it is hoped to use diagnostically, the path of tissues between the scattering site and the

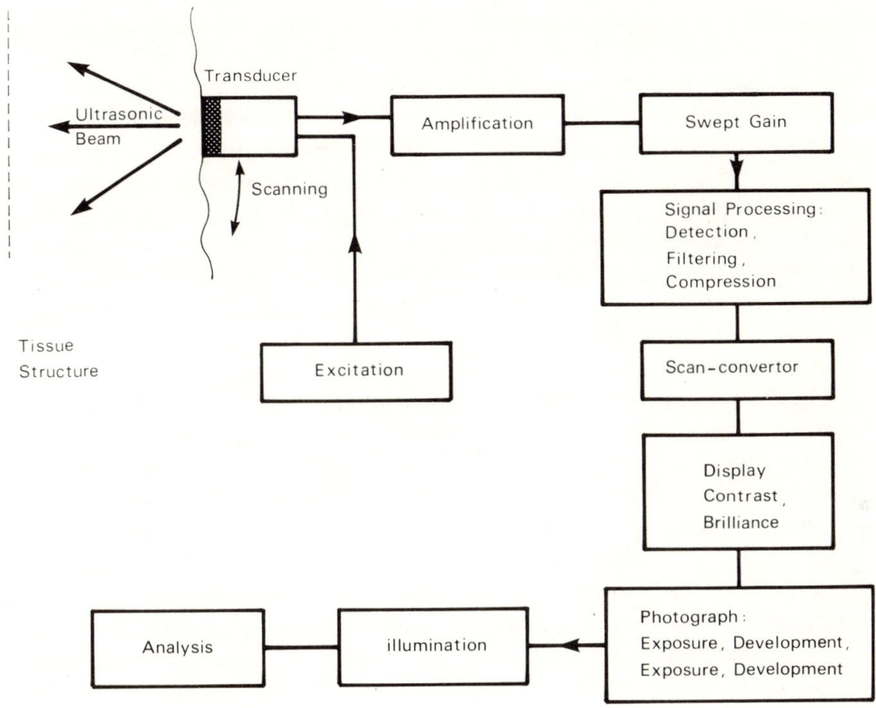

Figure 2.1 Schematic diagram of an ultrasonic pulse-echo scanner, indicating the parameters that may contribute to the final image. After (14).

transducer may be of profound importance. The relevant acoustical parameters appear to be velocity and attenuation. Not only is the path material inhomogeneous in the sense that scattered echoes (presumably caused by inhomogeneities) seem to originate from nearly all soft tissues, but the path is piecewise inhomogeneous *in vivo*, with unknown thicknesses of tissues of unknown properties. The question of wave propagation in these tissues and the choice of an appropriate model using currently available experimental data is discussed in section 2.2 below.

The final complexity is introduced by the transducer element itself. On the basis of the simple interface model, the linear dimension of the element is kept as small as possible to improve lateral resolution, while still maintaining significant directionality. In the low megahertz frequency range (1–10 MHz) usually used in diagnostic visualization ($0.15\,\text{mm} < \lambda < 1.5\,\text{mm}$), this leads to highly diffractive fields which further complicate a rigorous approach to the analysis (26). It is only relatively recently that a careful and successful comparison (27) has been made of the phase and amplitude distributions radiated by a transducer element in practice, with those predicted for an ideal transducer. The introduction of effective geometrical parameters (28) for the radiator appears to be necessary to provide good agreement in many cases. Unfortunately the performance of a transducer element cannot be predicted *a priori*, and the field measurement techniques developed to date are limited to frequencies below about 4 MHz by technical considerations (29). It is worth noting in passing that the rectangular elements predominantly used in real-time scanners have potentially a much more complex field structure than the circular disc elements used in compound scanners (30). Even if the element functions in an ideal way in reception, it may still introduce significant problems in the quantification of scatter (see section 2.2 below).

The features of the electronics that may affect the image produced are several and reflect the technological emphasis in the development in ultrasonic instrumentation, in preference to an improved understanding of the scientific basis of the imaging process. The electronic features of the wide variety of commercial equipment available give the machines from different manufacturers their distinguishing characteristics. The form of the electrical impulse used to excite the transducer is likely to vary somewhat but, with a common tendency to strive for the shortest acoustical impulse possible in order to produce the best axial resolution, the transducer characteristics probably dominate the shape of the field finally radiated. More variation is seen on the processing of the received signals. There is a tendency to distinguish between electronic processing that is implemented prior to the deposition of the image on a temporary storage device (scan convertor)(31), and that which is carried out subsequently, often with the aid of a computer. These are called 'pre-processing' and 'post-processing' respectively (32).

The echo signals detected by the transducer are first subjected to broadband amplification and then to some form of time-dependent gain to compensate for the attenuation of the pulse as it travels deeper into the body. The time-gain compensation is achieved in a number of ways, usually involving at least three controls and giving an element of piecewise adjustment. The r.f. signal is then demodulated and filtered to produce the video signal. Before or after the demodulation, the signals need to be subjected to some form of compression amplification (33) to try and accommodate the 60 dB of useful signal dynamic range available at this stage to the much lower dynamic ranges of the storage and display devices (34). The characteristics of the compression used also vary between manufacturers.

The scan convertor on which the image is stored may be analogue or digital. The performance of the analogue devices is controlled by video bandwidth, dynamic range, deflection slew time, registration and resolution. The performance of digital convertors is largely controlled by the manufacturer's design. Although the two types of devices have broadly similar characteristics (31, 32), the differences may be significant in affecting the image characteristics. The transfer of information to the display device introduces display variables of contrast, brilliance etc. It may also involve some further selection of information since the potential 40 dB of the store requires further accommodation to the 30 dB or so of the dynamic range of a good display device. Post-processing options are essentially limited by the imagination of the investigator and the computing facilities available. They include spatial Fourier transformation (35), grey scale windowing, weighted averaging, and full frame histogram equalization (32).

If perceptual studies are to be performed on a series of images, these must be recorded in some permanent form. For ease of testing with different observers a series of photographic records is frequently used. This usually implies a further loss of dynamic range and introduces the further variables of the photographic processes and the illumination of the final image. If consistent results are to be obtained it is necessary to control both these and the performance of the display device (36).

From the brief summary given above it is clear that optimization of the diagnostic imaging process requires both a reduction in the variables that enter the process, and also careful control of those that remain. Considerable efforts have been devoted to the various stages of the process of ultrasonic grey-scale imaging—by experimental model studies, by computational and analytical modelling, and by psychoperceptual studies. Unfortunately the work described is seldom put in any broader context than that of the particular area with which it is concerned. As an almost independent endeavour at the present time attention is being focused on empirical approaches to provide quantitative indices for differential diagnosis in specific clinical situations (37–39). The quantitative nature of the approaches removes the problem of operator transferability, while leaving untouched the problem of transfering the techniques from one machine or laboratory to another. The full exploitation of these techniques is likely to need prior solution of this latter problem. It is very probable that this will in turn require the same understanding as is required for the optimization of the basic imaging systems (although it may be differently applied). The critical factor is an understanding of the basic mechanisms which give rise to the scattered echoes that are being visualized.

2.2 Acoustical wave propagation in tissues

Reference to figure 2.1 suggests that the measurement of the backscattering that is subsequently visualized in pulse-echo imaging can be achieved, even *in vitro*, only after determining the other propagation parameters, viz. the velocity and attenuation of the intervening tissue (21, 23, 40). At least in the laboratory the tissue need not be piecewise inhomogeneous and its properties may be determined by transmission measurement techniques. Considerable effort has been directed at measurement of the velocity and attenuation of ultrasound in soft tissues since the initial visualization and the results given in table 2.1 (8, 9, 41).

2.2.1 *Velocity*

From the data in (8, 9, 41) it can be seen that the initial assumption of an approximately constant velocity was a fair one, most of the data lying in the range 1500–1600 m/s. For the purposes of visualization the few per cent variation that this implies is probably not significant. For the purposes of dimensional measurement, as used in obstetrics and opthalmology, the range is significantly reduced by selecting a value of velocity from the much narrower range of measured values appropriate for the tissue whose dimension is being measured.

The range of values apparently obtained for velocity may in fact be influenced by other considerations. The volume of data available reduces markedly when it is categorized according to the animal used, the frequency of the investigation, the temperature of the measurements and the condition of the tissue specimen (9). The data tabulated have value, but need to be used with extreme care. The diffractive nature of the piezoelectric transducers commonly employed for the measurements, which was mentioned in the previous section, demands that a correction be made if the non-planar wave fronts that are perforce used in practice are not to influence the values obtained (42). It has been pointed out (23) that the procedure of diffraction correction for velocity is an essentially iterative one, since the velocity which is to be measured is needed in the calculation of the correction. The point at which the iteration stops is clearly determined by the numerical significance of the correction. For ideal transducers, the diffraction corrections for velocity need only affect the results obtained by a fraction of one percent. Not only may the use of non-ideal transducers affect the corrections adversely by a factor of two, but any attenuation of the medium also needs to be included in the velocity correction calculation (43). The correction analyses are based on the experimental situation in which the specimen whose properties are to be measured occupies the whole of the volume between the transducers. There appears to have been no attempt as yet to analyse the more common situation in which relative measurements are made before and after the specimen has replaced a (parallel-sided) region of the liquid (usually water) between the transducers.

The comparison of velocity data from one laboratory to another is thus hindered by the overwhelming tendency of authors reporting data to avoid consideration of the accuracy (as opposed to the reproducibility) of their measurements. There is a paucity of evidence for velocity dispersion in biological tissues (44). The apparent absence of dispersion commonly reported is inconsistent with the predictions of relatively conventional theoretical models which seek to describe the strong frequency dependence of ultrasonic attenuation in tissue (45–47). Recently Gurumurthy and Arthur have shown (48) that for a Hilbert model of propagation the velocity dispersion is likely to be small and thus requires very careful measurement. The accuracy needed is likely to be comparable with the magnitude of the diffraction corrections. Such measurements may be of considerable value in establishing the validity of a particular propagation model of this type.

2.2.2 *Attenuation and absorption*

The attenuation of soft mammalian tissues—which is of the order of 100 dB/m/MHz (8, 41)—is a gross feature of the wave propagation which is inescapable. Its accurate measurement and interpretation are not so obviously seen. In 1952 Fry (49) specifically discounted the possibility of scattering processes contributing to the 'absorption' of ultrasound in tissue. In 1968 Hill (50) distinguished between 'absorption' and 'true absorption', the present author suggesting the use of the terminology used in other fields of 'attenuation' and 'absorption' respectively (21). For a period the two terms were used interchangeably in the literature. Whether the scattering processes are sufficiently strong in soft tissues to contribute to the attenuation coefficient in a significant way, they were and are sufficiently strong to provide information for display and analysis in grey-scale imaging!

Some of the confusion in this context may be identified by consideration of the methods used for measuring ultrasonic attenuation. Traditional methods are based on an arrangement similar to that used for the measurement of velocity. The attenuation is derived from relative amplitude measurements made before and after the specimen has displaced a section of the liquid between the transducers. The vast majority of the measurements reported in the literature (8, 41) have been made using finite piezoelectric transducers. The implicit model behind this measurement approach is that of plane wave transmission (i.e. a one-dimensional situation). Some concession to reality can be (although seldom is) made by the introduction of diffraction corrections as mentioned in the previous section. Conventionally the diffraction corrections for attenuation measurements only require prior knowledge of the velocity (42) (and the effective dimensions of the transducing elements). A more rigorous computation (43) shows that both the velocity and the attenuation are needed and the correction procedure, as for velocity, becomes an iterative process that stops at the point of numerical insignificance. Diffraction corrections for attenuation are typically an order of magnitude (as a percentage) greater than those for velocity.

The feature of the piezoelectric receivers that makes diffraction corrections necessary also introduces another complication. In reception the voltage output as a result of the transduction process is proportional to the complex pressure amplitude averaged over the surface of the transducing element. Thus if, for some reason, the wave impinging on one half of the element has the same amplitude as that impinging on the other half of the element, but is out of phase with it, zero voltage will be output from the transducer. The immediate implication of extremely high attenuation is misleading. This mechanism, which can in principle produce unlimited systematic errors in the measurement of attenuation in heterogeneous media, was identified by Southgate (51), and subsequently by Marcus and Carstensen (52). It has become known as the 'phase cancellation artefact'. It has been shown, by reducing the size of the receiver and by using energy dependent detectors such as calorimeters and radiation force devices (53), that typical errors in measurements on soft tissues may be several hundred percent (e.g. (54)). Such errors may be present in much of the data on attenuation reported in the literature (8, 41), and will thus have a significant effect on efforts to identify valid propagation models (45–48).

The existence of this artefact for certain specimens of tissue is fairly well documented. The identification of its source and its significance is not. Phase fluctuations in a wave are almost always accompanied by amplitude fluctuations (55). If these fluctuations arise from the scattering of the incident waves by the inhomogeneities in the tissue, then a more sophisticated analysis is required. The amplitude and phase of the waves emerging from the specimen fluctuating across the specimen will lead to fluctuations in the apparent attenuation and velocity measured. The simple one-dimensional model implicit in the measurement procedure may no longer be appropriate. If not, multiple scattering and concepts of coherent and incoherent scattering (56) need to be introduced and related to the measurement procedure. Before this is attempted, the origin of the phase and amplitude fluctuations in the waves emerging from the specimens needs identification. Whereas results from energy dependent (phase-insensitive) detectors give evidence for the existence of the fluctuations, they cannot delineate the mechanisms of their production.

Based on carefully calibrated miniature hydrophones (57, 58), Aindow *et al.* have developed procedures (59–61) for measuring the phase and amplitude distributions of the waves emerging from tissue models (see section 2.4 below) and tissues. The fluctuations observed by earlier workers could have arisen from one or more of three sources: the use of non-planar incident waves, variations in the specimen thickness, and from scattering by the internal inhomogeneity of the tissue. Removal of the first two of these by careful experimental technique indicated that heterogeneities can produce significant fluctuations, but the degree to which they are induced varies considerably between different specimens of the same tissue in apparently the same condition (62).

Thus it appears that soft tissues may contain sufficient inhomogeneity to invalidate the essential principle on which the measurement procedures have been interpreted—the one-dimensional model in which plane waves continue to propagate as plane waves through the tissue or in which straight line ray propagation may be assumed. Very considerable further work is needed to assess the numerical significance of this effect for different tissue architectures, compositions and conditions. The immediate consequence is that the definition of attenuation conventionally used applies only to those situations in which the scattering is sufficiently small that it does not significantly disrupt the wavefronts. In such circumstances attenuation and absorption may well be indistinguishable within the experimental errors of the measurements. Bamber (63) has detailed typical error contributions for velocity, attenuation and scattering measurements. He gives as the greatest single source of error in the attenuation results, (apart from phase-cancellation effects), the uncertainty in the measurement of the thickness of the sample, which is typically 10%. There appear to be few deliberate attempts to separate the absorption and the attenuation. Goss *et al.* (64) comment that the relative contribution of the absorption to the attenuation depends on the method used to measure attenuation. Unfortunately these authors did not discuss scattering in their analysis.

2.2.3 *Scattering cross-sections*

There have been a number of theoretical analyses of the scattering of ultrasound by soft tissues (e.g. 21, 56, 65, 66). These have almost all used an incident plane wave, and derived an expression for the scattered wave in terms of an intensity or in terms of the mean square of the pressure amplitude. The Born approximation (weak scattering) is usually invoked and the media are assumed to be lossless. The inhomogeneities in the media are described by continuous parameter variations. The parameters that are used are the velocity, the velocity and the density (which are of course related), or the compressibility and the density. The inhomogeneities are sometimes said to be due to variations in the characteristic acoustic impedance (which is the product of the velocity and the density), following the ideas present in the plane interface model. However this tends to oversimplify the essential physics of the situation. The directional properties of these different scattering

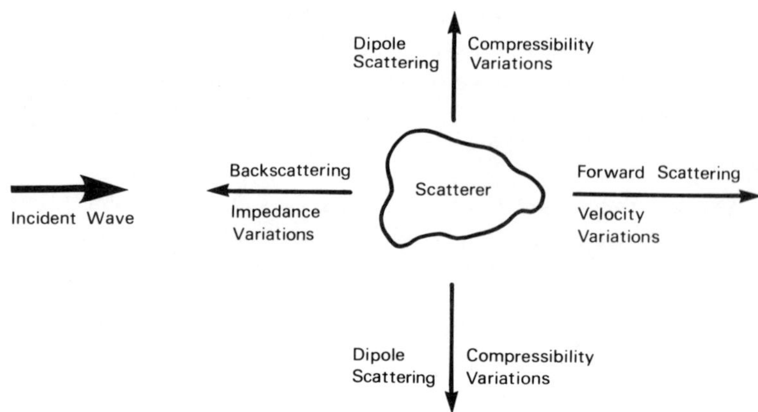

Figure 2.2 Directionality of scattering from parameter inhomogeneities, assuming weak scattering and an absence of absorption scattering. After (67).

contributions if the scattering is weak and that absorptive scattering can be neglected is shown in figure 2.2 (after (67)).

Relating these theoretical analyses to experimental situations is relatively difficult except in rather general terms. For the purpose of interpreting measurements, a number of authors have developed specific analyses that permit the backscattering cross-section per unit volume to be calculated from measurement results. Probably the most comprehensive discussion of the problems is that of Reid (40), who also reminds us of the necessity of knowing the velocity in and attenuation of the tissue sample under investigation before the backscattering cross-section can be evaluated.

In the light of the dicussion of the previous section (2.2.2), this apparently simple statement places implicit preconditions on the nature of the scattering that is present and that can be experimentally analysed. An assumption comparable to that of straight line ray propagation enters all the scattering analyses that have appeared (40, 63, 66, 68, 69). Furthermore the analyses derive expressions for scattered intensities, thus specifically avoiding the problem of phase cancellation artefacts for scattering measurements. The procedure of averaging the measurements over a number of specimens or scattering volumes within one specimen (e.g. 66, 69, 70, 71) does not, of course, remove the effects of phase cancellation which can only affect the results in one direction (giving lower values for the scattering than may be realistic). By working with thin samples and in the far field of the scattered waves, it is possible to overcome the majority of these problems, and while this can be achieved in the laboratory, with some care (71), the situation *in vivo* does not permit this flexibility. The potential effects of phase cancellation on scattering measurements have been subject to computational investigation

(72). The computational model assumes straight-line ray propagation from the scattering centres to the receiving transducers.

If, as the preliminary work of Aindow *et al.* discussed in the previous section suggests, the assumption of straight line ray propagation in the tissue is invalid in some cases, then it is difficult to see how meaningful quantitative measurements can be achieved in these circumstances. Even if a plane wave is incident on a tissue specimen in the laboratory, the scattering between the specimen boundary and the volume whose scattering is to be measured will introduce phase and amplitude fluctuations across the wave front that

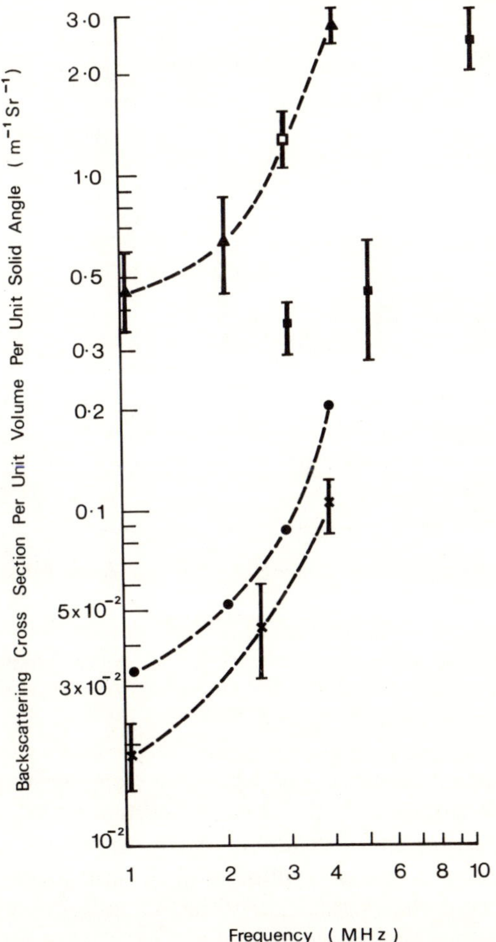

Figure 2.3 Backscattering cross sections for normal liver: ▲ fresh human (75); ● fresh human (76); ■ fresh calf, □ fixed calf (77); * fresh human (78).

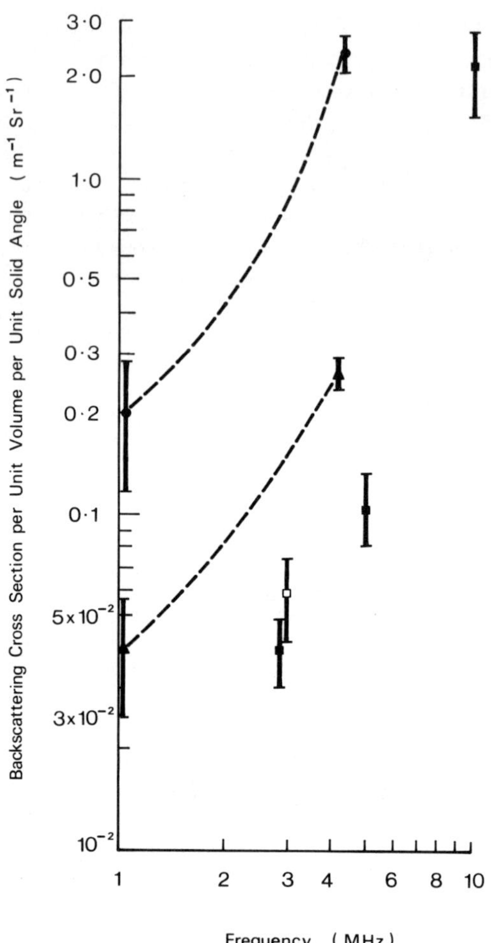

Figure 2.4 Backscattering cross sections for normal tissues: ▲ fresh human brain (75); ● fresh calf myocardium, □ fixed calf myocardium (77); ● fresh human spleen (75).

interrogates the target volume. The scattering from the target volume will impose a fresh distribution on the wavefronts that set off to return to the transducer; but this distribution will be further moderated by the scattering of the tissue between the target volume and the transducer. The output of the transducer is the complex resultant of the fluctuations that finally emerge. The separation of the three elements of fluctuation present (two intervening tissue paths, and the scattering itself) would appear possible only if the intervening path fluctuations are insignificant compared to the scattering fluctuations. This will depend on the tissue specimen under investigation and the depth of the target volume into the tissue, since fluctuations tend to increase with increasing path length (55). The effect of strong attenuation in reducing the path length dependence of the fluctuations has yet to be analysed.

In view of the above it is not surprising that the experimental data on backscattering cross sections from tissue are rather limited. The excellent agreement of the experimental results on blood with a theoretical model (73, 74) reflects two factors. Firstly that the scatterers (viz. the red blood cells) and their properties are relatively well defined. Secondly, it reflects the fact that the level of scattering produced is quite small and thus the assumptions of the theoretical analysis applied (primarily the Born approximation) are valid. Data available on other tissues are shown in figures 2.3 and 2.4 (75–78), with the exception of that included in reference (79) which shows the relationships between backscattering cross section per unit volume per steradian and the fat, water and collagen content of the tissues. It appears from figure 2.3 that a factor of 30 exists between the results reported for normal liver at 4 MHz. This variation is not inconsistent with the hypothesis that the assumptions of the analyses used are not always valid, as has been suggested above.

The identity of the scatterers (and thus their dimensions and properties) remains ill-defined, although there have been continuing speculations based on histological information. Nicholas (76) has suggested a bimodal scatterer distribution. The suggestion is based on fitting backscattering data to a theoretical model in which density variations in the tissue are neglected compared to compressibility variations. The primary reference for this assumption is the assertion of Fields and Dunn (80) that the density of soft tissues varies by less than one per cent. The data of Wells (7) do not support this assertion, even at the level of the bulk properties which, being measured over relatively large volumes, are inappropriate for the scattering problem under discussion which is concerned with much smaller dimensions. Unfortunately data in the appropriate range are extremely scarce, although those available on cellular fractions (81) confirm the inaccuracy of the assumption. It is relevant that the dipole (density) term was used in the successful analysis of scattering from blood (74). Nicholas suggests typical dimensions of approximately 20 μm and 1 mm for the scatterers. Whereas the scattering theory and calculated results may well be valid for the former (depending on their scattering strength), it is not clear that they would be valid for the latter. This is of importance in considering the modelling of the imaging process that has been performed.

2.3 Modelling of the imaging process

There has been relatively little work in this area, although both analytical and computational models have been reported.

2.3.1 *Analytical modelling*

The first contribution to the problem appears to have been the indirect one of Atkinson and Berry (82) who were concerned to analyse the fluctuations of amplitude appearing in the signal scattered back from blood. With appropriate assumptions for the scattering from and the propagation of ultrasound in blood, they provided an analysis which showed fair agreement with experiment. The envelope of the fluctuations observed was, for a dense distribution of small weak scatterers, shown to be due largely to the pulse shape employed.

A more general analysis, specifically of the pulse-echo imaging problem, has been given by Gore and Leeman (83). With the assumption of very weak scattering (and thus no multiple scattering) they showed that the signal arriving at a point receiver contains not only information on the microstructure of the tissue, but also a pulse smoothed version of the microstructure. As in Atkinson and Berry's analysis, they predict that the use of a transducer operating at the same frequency, but with a different beam profile, would produce a different signal from the same tissue structures. Both of these analyses assumed a lossless medium although Gore and Leeman indicated that the extension to include absorption would cause no problems (83).

Independently of this work, Burckhardt has produced an analysis (84) starting with a random distribution of scattering phases and amplitudes. The analysis has similar (implicit) assumptions to the earlier work and reaches similar conclusions. An analogy is drawn between the speckled appearance produced in ultrasonic B-scans and laser speckle, although the author indicates that the validity of such an analogy remains an open question. The analysis is applied to compound scanning regimes which are shown to give an improvement in the signal to noise ratio whether maximum amplitude writing or amplitude averaging is used. Burckhardt predicts that displacement of the transducer by an amount at least equal to its width is needed if an independent set of amplitudes is to be obtained in the image. The speckle analysis indicates a compromise between resolution and signal to noise ratio in that large-aperture, high-resolution probes have a lower signal to noise ratio than lower resolution ones. The speckle approach has been extended somewhat by Abbott and Thurstone (85). They have shown that the speckle is caused primarily by the phase fluctuations across the face of the receiving transducer rather than from amplitude fluctuations. An experimental investigation of Burckhardt's prediction on the distance that the transducer had to be moved to obtain an independent sample gave good support for the theory. The procedure used was to take a number of pictures at different transducer locations, and to ask the observers whether they considered the pictures, taken in pairs, 'very alike', 'somewhat alike', or 'very unlike'.

The extensive analysis of Fatemi and Kak (86) includes a discussion on scattering from inhomogeneities which is very similar to those described above (21, 56, 65, 66). These authors include the effects of absorption in the medium by the formal introduction of a complex wave number, but the theory is not developed further. A point spread function of image degradation is derived for linearly scanned unfocused transducers. Whereas they show that under certain circumstances this may be made field position invariant, their results are very limited and are concerned only with dual wire targets.

2.3.2 *Computational modelling*

The complex process of image formation shown in figure 2.1 has been the subject of a detailed computational modelling (87) in which both the tissue structure and the imaging system is modelled. The simulation explicitly assumes an absence of attenuation and noise, as well as ideal instrumental characteristics as far as amplification and transducer behaviour are concerned. The transmitted sound pulse is described as a sinusoidal r.f. pulse modulated by exponential leading and trailing edges. Its lateral distribution is assumed to be gaussian. Signal processing options included are different modes of detection (half wave, full wave, or square law), and video smoothing (described by the time constant of a single pole filter). The tissue is assumed to contain variations in the compressibility only (neglecting density variations), and an absence of multiple scattering. The parameter used to describe the dimensional scale of the inhomogeneities is the correlation length (56).

The preliminary results reported indicate that if a tissue model with a large correlation length is scanned with a small pulse, the image obtained may contain more detail than the original model due to the complexity of the pulsed field distribution for short pulses (88). Full wave detection of the r.f. signal may also introduce spatial frequencies twice as great as the highest originally present. When the correlation length is much smaller than the pulse size (the situation assumed in the theoretical models mentioned above (82–85), the appearance of the image is, as predicted, highly dependent on the pulse length and beam width, the latter having some influence on the image structure in the axial direction.

A report more concerned with the effect of electronic variables on the image presented is that of Jones and Kimme-Smith (89). The parameters subjected to computer simulation include grey-scale resolution and image size, rectification and envelope detection processes, digitization rate, preprocessing algorithms, scan convertor storage algorithms, and post-processing algorithms. Some typical preliminary results are presented and the communication identifies the directions of future work as a more quantitative measure of the image texture displayed, and the comparison of such textural parameters with the visual evaluation of the same image. Sections 2.5 and 2.6 below describe the work that has already been reported in these areas. If the visual evaluation is to be performed using photographs, it is important that these are taken without using the frame freeze facility available on most displays, since this tends to maximize the speckle noise on the displayed image (90). In addition any still image display may in fact be a 'worst case' representation of the clinical diagnostic situation where the eye-brain system tends to filter out some of the more rapidly varying speckle (90).

It is clear that the analytical and computational results presented in this section are in general agreement with the conclusion that a distribution of scatterers that are sufficiently weakly scattering to exclude multiple scattering and to permit the valid assumption of straight line ray propagation, and which are sufficiently numerous for there to be many scatterers per pulse length, will produce a granular or speckled image in which the features are as much determined by the beam characteristics as by the microstructure of the tissue. Revealing as these models are, their potential value stems from the appropriateness of the basic assumptions on which they rest. As was discussed in section 2.2 it is by no means clear that these assumptions are in fact valid for all soft tissues. A line of investigation parallel to that of theoretical models is that of experimental models, which are discussed in the next section.

2.4 Experimental tissue modelling

The original conception of tissue modelling in ultrasonics (21) was to provide controlled systems of known parameters that could be used to gain some physical insight into the processes of scattering in tissue. Almost any inhomogeneous material that supports ultrasonic wave propagation will produce diffuse images when scanned with a commercial scanner; one of the popular materials to be used is natural sponge (87, 89). The main difficulty with these is the one we have in connection with human tissue, namely that of quantitatively determining the microstructure of the material in a non-destructive way. More controlled systems are needed and suspension systems tend to be most often used because of their flexibility. With independent variation of matrix properties, scatterer properties, and scatterer size and concentration, a systematic matching of acoustic parameters may be attempted (91, 92). Liquid suspensions (93) suffer from thermal sensitivity and instability as far as the distribution of scatterers is concerned. Interest has therefore centred on gel-type suspending matrices, although the very early use of blood as a source of standard echoes (94) has immediate appeal in the *in-vivo* situation.

The variety of materials used has been considerable. Initially the interest centred on producing a material that gave standard if arbitrary echoes as a means of equipment calibration. Till and Ossoinig (95) followed the author (21) in using glass beads suspended in silicone rubber, and were able to obtain quite reproducible sensitivity settings on their machines. Sommer *et al.* (96) used cellulose fibres and plastic beads in gelatin for the testing of scanners to be used for the evaluation of fluid-filled masses. The quantitative approach to tissue modelling was revived after a period of five years by Madsen and his colleagues (97). Using a suspension of graphite particles in gelatin, as described by Kol'tsova and Mikhailov in 1970 (98), materials with appropriate velocity and attenuation were developed. These were followed by reports on oil: polymer mixtures (99), cellulose fibres and glass microspheres in gelatin in an attempt to model the human torso (100), dextran spheres in hydrocolloid gels (101), graphite particles suspended in agar (102), reticulated foam pads immersed in liquids (103), and oil droplets suspended in gelatin (104). In the majority of cases (97–102, 104) careful measurements have been reported testifying to the concern to ensure that the medium has an ultrasonic velocity typical of that for soft tissues, and approximately the correct frequency dependence of attenuation. This latter has been achieved on the basis of the data available in the literature on the values of attenuation in soft tissues. The degree of tissue equivalence achieved as far as this parameter is concerned is only as good as the data available for matching, and the reservations on the data in the literature discussed in section 2.2.2 are relevant in this context. The importance of matching the velocity and attenuation of the tissue models as far as machine assessment is concerned lies in the fact that the former is needed for considerations of registration if compound scanning or focused transducers are used, while the latter removes the necessity for changing the swept gain controls from the settings regularly used in clinical practice.

The use of these phantoms is extremely limited at the present time. They may be used to simulate tissue in a fairly crude way while primary attention is being devoted to the measurement of scanner resolution using regularly spaced wires or nylon strings. They may also incorporate a series of graded visibility features for assessing beam shape and compression characteristics (105, 106). Until very recently no attempt had been made to make the phantoms tissue-equivalent as far as the parameter that is visualized in the texture is concerned, namely their backscattering characteristics. The recent first attempt to remedy this (107) raises the fundamental question outlined in section 2.2.3 as to the nature of the backscattering that does arise from soft tissues. The graphite: gel suspensions currently most common in use almost certainly have scattering properties that fulfil the assumptions of the scattering measurement analyses of section 2.2.3, and also of the analytical and computational models of section 2.3. There is as yet no evidence that they produce scattering images in the same way as soft tissues.

This raises the other fundamental weakness of such models for scanner assessment at the present time, namely that the assessment of grey-scaling and texture is essentially a subjective one, as is the comparison of the textures produced by models and tissues. Even if the scattering properties of tissues were well defined, and models constructed to simulate them, the models would be used in a relatively uncontrolled manner unless the images were quantified in some way. Shortly before the first commercial scanner provided a facility for a quantitative analysis of a selected area in an image, (in terms of a simple histogram of the fraction of the selected area at a given brightness level), a pilot study was reported (108) in which a similar analysis was performed on photographic records of clinical scans using a commercial image analyser. The detailed discussion of this paper defined an important role for quantitative analyses of the images, elements of which have been introduced in previous sections.

The rigorous definition of the mechanisms of ultrasonic scattering in human soft tissues and the identification of the microstructural features of the tissues which are involved, and the relation of these to disease processes and visualization procedures, clearly require considerable further work. When it is achieved it will still be necessary to consider the optimization of the texture displayed for visual discrimination. The contribution towards this that can be made by the use of quantitative indices of the images produced is that they can act as an objective, if arbitrary, bridge between subjective perception on one hand and instrumental variables on the other. Thus they may be used to provide a set of measures, and possibly even a language basis, for the displayed images which are to be optimized for perceptual discrimination; and at the same time the sensitivity of these indices to variations in such factors as model composition and properties, transducer beam patterns, and signal processing can be assessed.

The remainder of this chapter is devoted to preliminary investigations in two of these areas. The first (section 2.5) concerns a quantitative image survey of a number of commercial scanners using a series of tissue models, while the second (section 2.6) describes the subjective assessment of a series of images produced by commercial scanners.

2.5 Quantitative image survey of commercial scanners

The models used for the studies described in the next two sections consisted of suspensions of glass beads in silicone rubber. The procedures for making the models have been described previously (109), as have the procedures used for assessing the evenness of the distribution and actual scatterer concentration achieved. Apart from the need to minimize the inclusion of microbubbles of air which can be quite effective in producing spurious textural

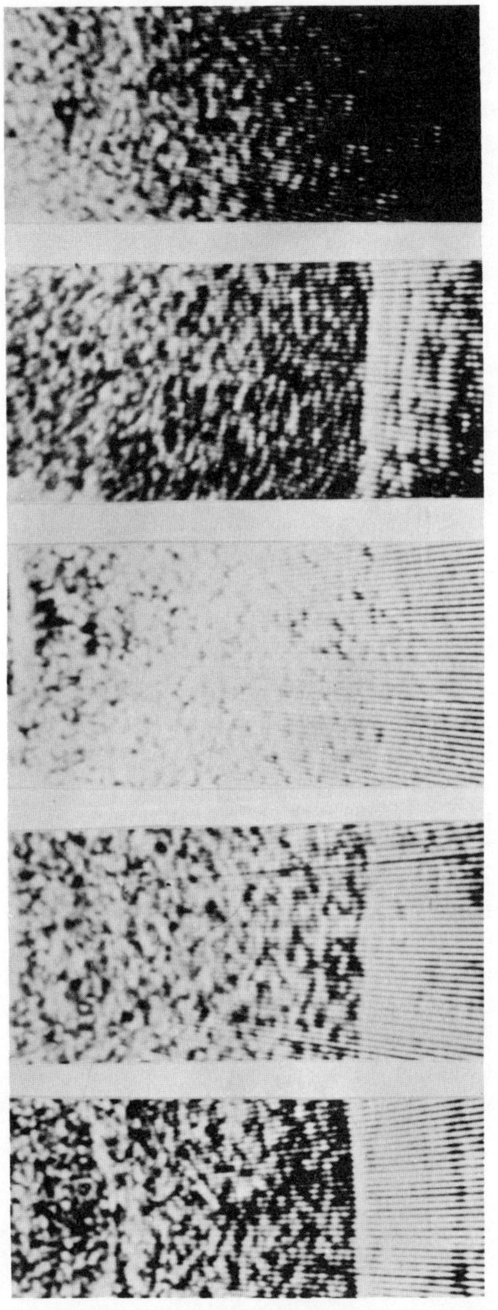

Figure 2.5 Commercial scanner visualization (at 2.5 MHz) of a series of models with 250 μm diameter scatterers. From left to right the concentrations by weight are: 0.5%, 1.0%, 2.0%, 10% and 20%. After (110).

details, uniformity of concentration is the initial target in suspension construction. There is no obvious way of assessing this independently for the graphite suspensions often used. For this reason a transparent matrix material and translucent scatterers were used in the present work. The silicone rubber has the additional advantages of an extremely long shelf life and good thermal stability. The use of glass ballotini for the scatterers has the disadvantage of having relatively strong scatterers, but permits the construction of models with a wide range of scatterer size, and thus scattering regimes.

The concept of scattering regimes can be seen illustrated in figure 2.5. This shows the images obtained from a commercial scanner set for clinical examination. The different images were formed by increasing the scatterer concentration (keeping the bead size constant at approximately 250 μm). The frequency of the scanner transducer was nominally 2.5 MHz. The detailed interpretation of these images (110) suggests that for the first three images (from the left) there is virtually a one to one correlation between the bright dots in the image and the scatterers in the model. Between concentrations of 2% and 10% by weight, the mechanism of image formation appears to have changed. The bright regions of the image probably correspond to variations in the localized distribution of the scatterers. The change from one regime to the other occurs over the range in which the pulse length becomes longer than the mean free path (scatterer spacing) in the model. Thus the scattering process at the higher concentrations approaches that of the speckle analyses described in previous sections. A rigorous identification with the theoretical models of section 2.3.1 is not possible since the assumptions of the analyses are not valid for these models in which the scattering is not weak and multiple scattering is almost certainly present. The loss of detail at the bottom of the greatest concentration is due to the fact that the attenuation in the model far exceeded the compensation set in the swept gain of the scanner. The similarity of the images for concentrations of 0.5% and 10% by weight emphasizes the importance of defining the scattering mechanisms in tissues in some detail. Until this is done there are likely to be severe restrictions on the value of the quantitative analysis of textured images from both tissues and tissue models.

Following a series of pilot studies on commercial scanners (whose results included those shown in figure 2.5) a series of models involving five concentrations of four scatterer sizes were constructed. The concentrations were defined in terms of the number of scatterers per unit volume in view of the potential importance of the pulse length in the imaging process. The details of the models used are given in table 2.2. Advantage was taken of a commercial exhibition to obtain scans on the seventeen models of five different commercial scanners, all of which were real time devices. The probes were coupled to the models using proprietary coupling gel and the images

Table 2.2 Details of glass ballotini: silicone rubber models used for quantitative scanner assessment.

Model number	*Bead size range* (*μm diameter*)	*% glass by weight*	*No. of balls/kg of rubber* ($\times 10^{-5}$)
1	–	0	0
2	500–600	2.0	0.71
3	”	7.5	2.9
4	”	17.2	7.3
5	250–300	0.2	0.71
6	”	0.8	2.9
7	”	2.0	7.3
8	”	8.0	31
9	90–106	0.0092	0.71
10	”	0.037	2.9
11	”	0.095	7.3
12	”	0.40	31
13	”	5.0	480
14	53–63	0.0018	0.71
15	”	0.0075	2.9
16	”	0.019	7.3
17	”	0.08	31

obtained recorded by a time exposure on 35 mm negative film. The settings of the scanners were kept as they had been set for the exhibition. The camera settings were also kept constant throughout each set of exposures on a particular scanner. Subsequent development of the film and printing them at constant enlargement was performed with automatic processing to minimize potential variations from these sources. Photographs of the models and the scanned images obtained can be found elsewhere (111).

The image analyser used for the analysis of the images has been described previously (108). A light pen is used to select the area of interest, thus permitting the exclusion of obvious artefacts and boundary echoes from the investigation. A discriminating threshold of brightness can be manually adjusted, and one of the modes of operation of the instrument is the display of the area of the selected region which has a brightness above the preset level. The analyser has an automatic adjustment for the overall brightness of the total field of view. Thus although the photographic prints were all masked, the absolute values of the brightness recorded vary from one print to another depending on the strength of the scattering visualized. This variation has been removed by normalization of the brightness scales of the results using a linear interpolation between the extreme values which represent total darkness and total brightness respectively.

The raw data obtained, of the fraction of the area selected above a certain brightness level, has been converted to a more conventional display of the fraction of the area selected which is at a given brightness level. This is the presentation discussed in the previous section. The conversion was achieved by forward differencing the data on a point by point basis. Whereas this gives good results on a large number of closely spaced data points, it can give erratic results if there are relatively few points.

The results of the analysis for four of the scanners are shown in figures 2.6 to 2.9. The curves in each figure represent different concentrations of a particular bead size. The full results are discussed in detail elsewhere (111), and only the main features will be indicated here. Figure 2.6 gives the results for the largest bead sizes (500–600 μm). Two of the graphs include the model with no scatterers, model number 1, which clearly illustrates the problem of forward differencing on a very few data points. The similarity of the results for model 2 on machines *B*, *C* and *D* is encouraging. The consistent dissimilarity of model 2 results compared to those from the other concentrations of the same bead size is strongly suggestive of a change in the scattering regime between the two smaller concentrations. Apart from machine *A*, the distributions have a similar shape, even though the heights and positions of the peaks vary from one machine to another.

The use of models with smaller scatterers (250–300 μm, figure 2.7) shows two interesting features. There are clearly differences between the models, but the general form of the distributions tends to be more characteristic of the machines. The bilobed distribution of scanner *A* is almost certainly due to the fact that the display from this scanner has a marked line structure even when there is no ultrasonic image present. Thus the peaks at lower brightness should correspond to this line structure. The variation observed in them gives some measure of the overall imprecision of the analysis procedure used. The second feature of the distributions in figure 2.7 is the existence of a single distinct peak in the results from scanners *B* and *D*. While the general shapes of the distributions for these two scanners are the most similar, the models for which this distinct peak occurs are not the same in the two cases, suggesting that the scanner plays an important role in the distributions obtained.

Using even smaller scatterers in the models (90–106 μm in figure 2.8; 53–63 μm in figure 2.9) the apparent differentiation of the models increases. In part this may be due to the fact that the scattering from these beads is relatively low and the number of points along the brightness scale is thus reduced. It is noticeable that these scatterers are still very much stronger than the graphite particles commonly used, the strength of the scattering that the latter exhibit being thus due to the very high concentrations involved. It is interesting that the model with a uniquely high concentration in this study—model 13—does not tend to produce highly characteristic distributions (except in the case of scanner *B*). This result is almost certainly an artefact caused by the forward differencing routine used.

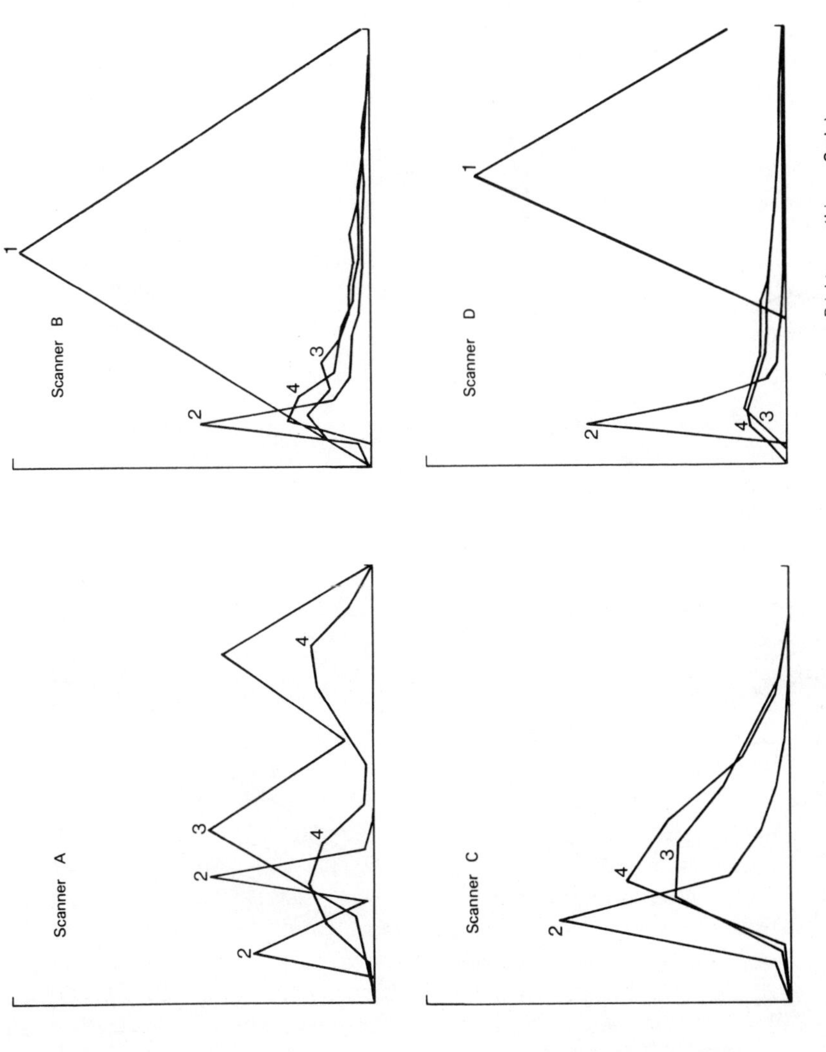

Figure 2.6 Quantitative analysis of models imaged on four commercial scanners. Bead size range: 500–600 μm. The curves are labelled according to the model numbers given in table 2.2.

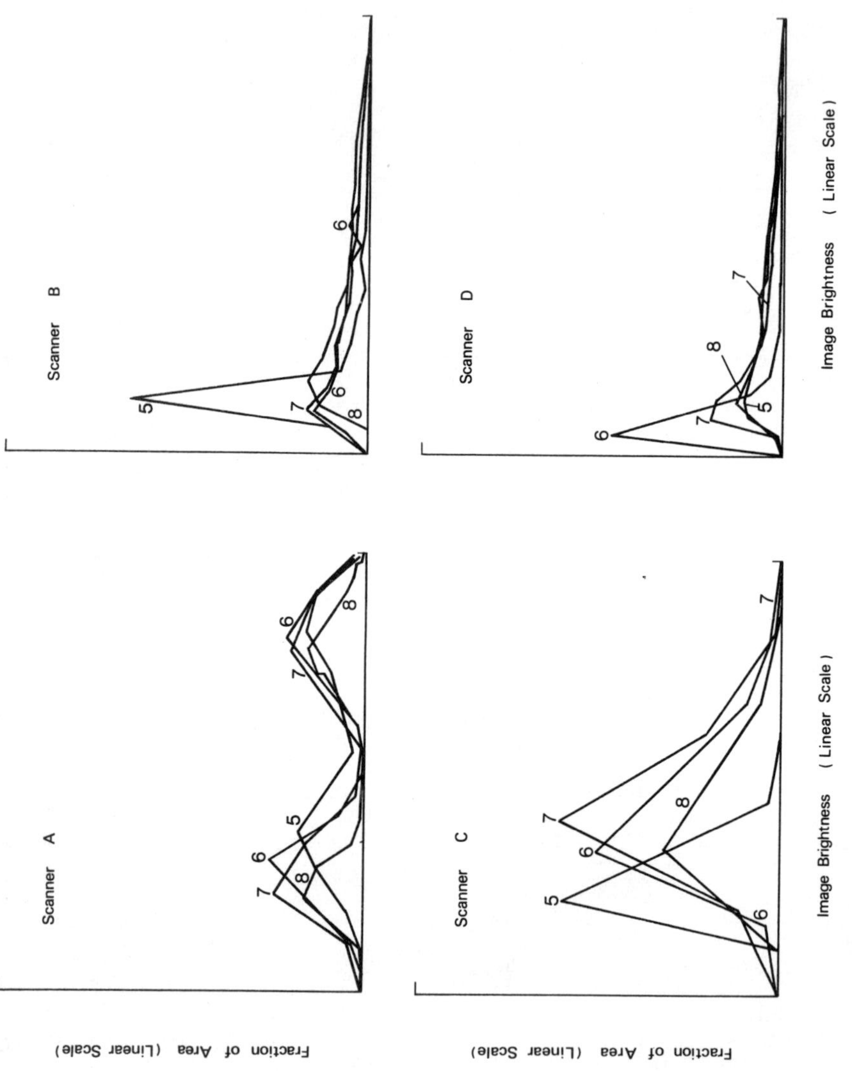

Figure 2.7 Quantitative analysis (see figure 2.6) for models with bead sizes 250–350 μm.

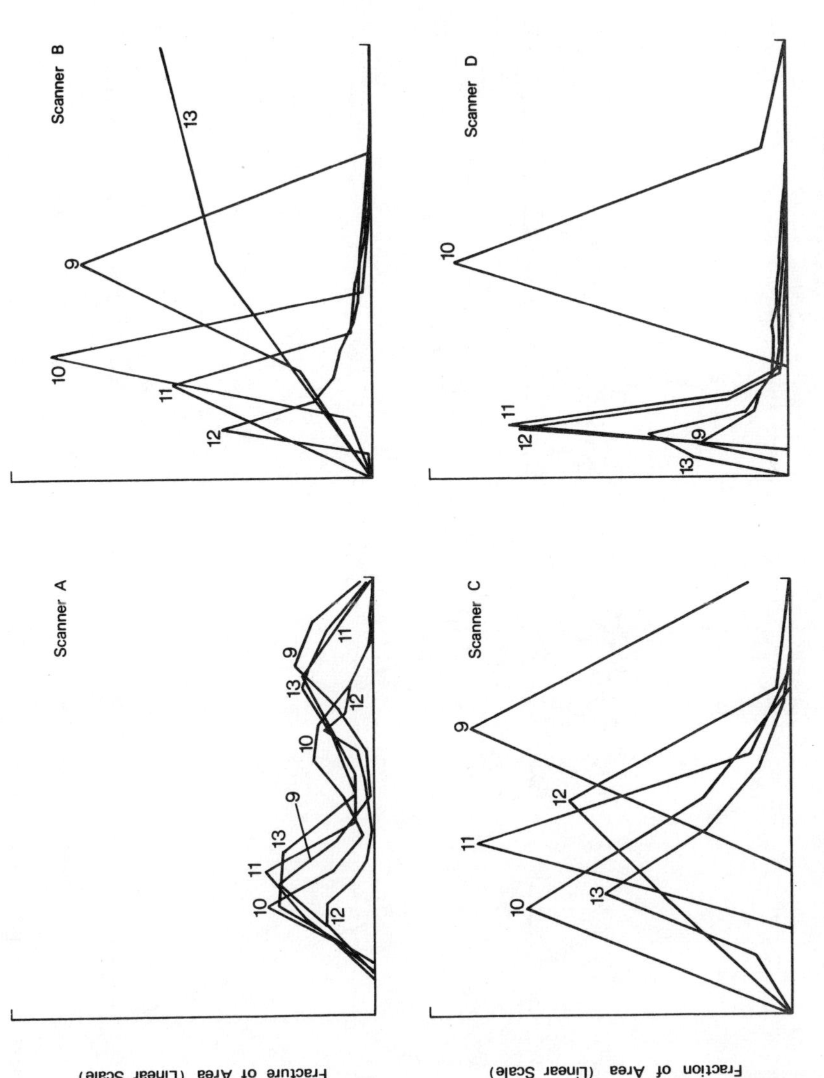

Figure 2.8 Quantitative analysis (see figure 2.6) for models with bead sizes 90–106 μm.

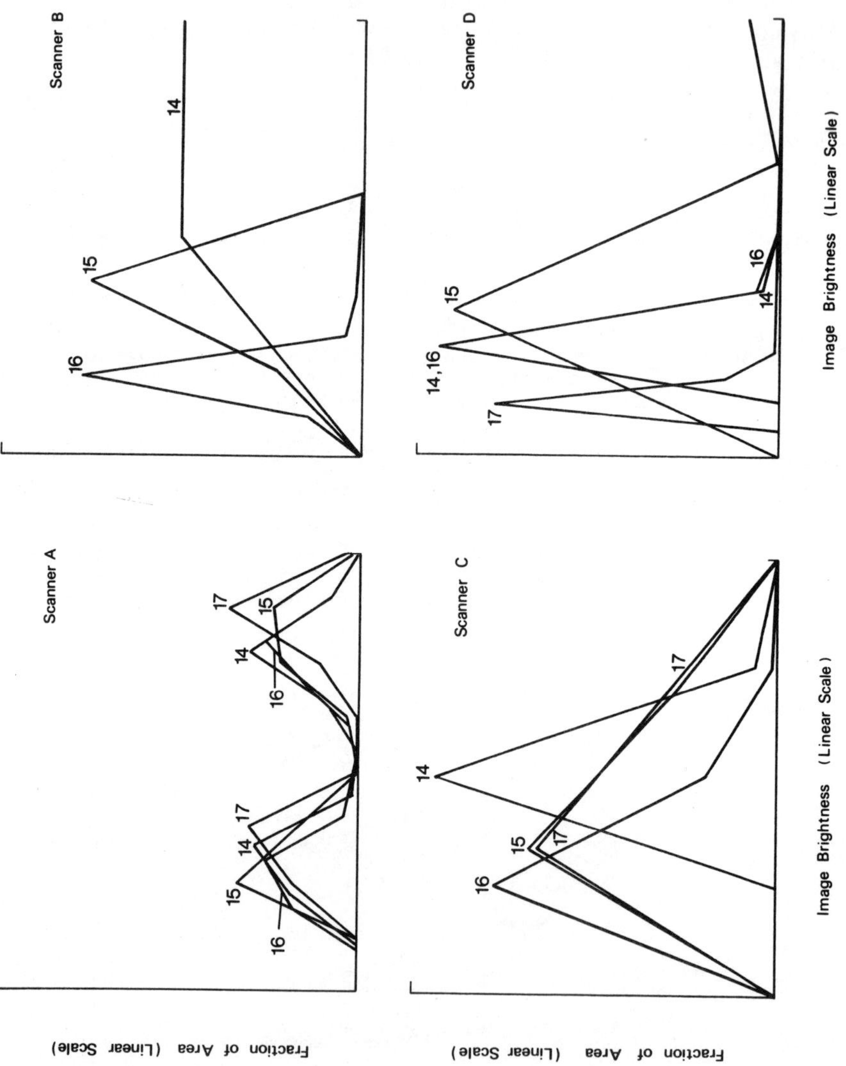

Figure 2.9 Quantitative analysis (see figure 2.6) for models with bead sizes 53–63 μm.

The survey outlined above has confirmed the crucial role that the scanner characteristics play in the structure of the images obtained. It is also clear that the use of image indices for empirical tissue characterization (37–39) is most effectively performed by operating on the signals as close to the output of the r.f. amplifier as possible. The use of brightness histogram facilities built into the scanner may well be valuable but difficult to transfer between scanners from different manufacturers. It is also wise to try and avoid the use of parameters that are strongly dependent on the multi-element gain functions of the scanner electronics (38).

2.6 Subjective assessment of scanner images

As a pilot study for the perceptual analysis of ultrasonic scans, four further models were constructed. The design of these is shown in figure 2.10. The concentrations used in these models are detailed in table 2.3. They were selected on the basis of the experience of the models described in the previous section as having visual characteristics, when scanned, covering a range of those observed when tissues are scanned *in vivo*. The models included wires for the assessment of resolution, but a discussion of the results obtained is outside the present context. Four commercial scanners (differing from those previously used) were adjusted by the same experienced operator for optimum visualization of the same subject's liver. Without changing the scanner settings, the four models were scanned and the results recorded on the hardware available on the scanner for the provision of hard copy. The resulting scans were then grouped according to the scanner used, and for each group five

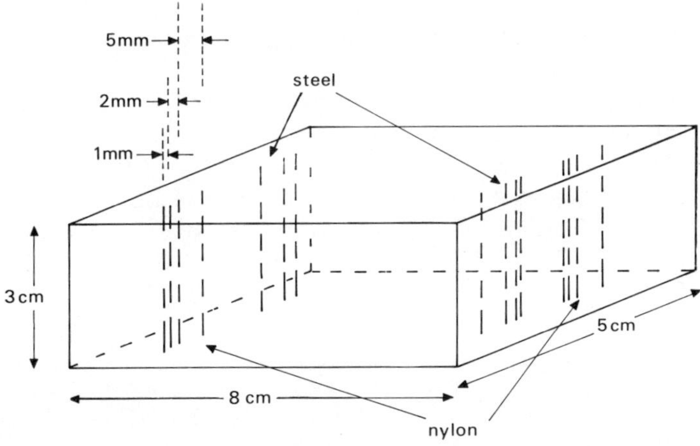

Figure 2.10 Construction of the models used for the visual assessments .

Table 2.3 Details of models used for visual assessment.

Model number	*Bead size range* (*μm diameter*)	*% glass by weight*	*No. of balls/kg of rubber* ($\times 10^{-5}$)
18	53–63	0.31	120
19	"	3.03	1200
20	90–106	1.54	120
21	"	13.5	1200

trained and one untrained observer(s) were asked to rank the model scans according to their similarity in texture to the liver scan on that particular machine. The results obtained are shown in table 2.4.

It can be seen that the trained observers were (probably fortuitously) unanimous, and differed only slightly from the untrained observer in their judgements. The detailed analysis of the results (112) features the tendency of the smaller scatterers to produce images more characteristic of tissue, and the fact that different models give patterns closer to liver texture on different machines. Thus considerable care needs to be used in assessing the performance of a number of machines with only one model. There is also an indication that different machines tend to emphasize different acoustic characteristics of the media scanned.

Following the pilot study, the same set of images were subjected to a more detailed perceptual assessment. This had two parts (113). The first was an extended version of the pilot study, while the second was concerned to see whether the perceived similarity between images generated from the set of models was consistent across different scanners. The technique developed by Shepard (114) of generating a matrix of similarity judgements of all possible pairs of images was employed in the hope that it might be possible to gain some insight into the dimensions of variation that the subjects used as their basis for discrimination.

The first part of the extended study was conducted with eleven male and eleven female unskilled volunteers in the age range 20–25. The procedure used was as before, the images within each set (from a given scanner) being presented to the subjects in random order. The results are shown in table 2.5 where it can be seen that the conclusions of the pilot study are confirmed quite

Table 2.4 Rank ordering of the model images in their similarity to liver texture (pilot study).

	Machine			
	E	*F*	*G*	*H*
Trained observers (5)	19, 18, 20, 21	18, 19, 20, 21	19, 18, 20, 21	18, 19, 20, 21
Untrained observer	19, 18, 20, 21	18, 19, 20, 21	19, 18, 20, 21	19, 18, 20, 21

Table 2.5 Results of the rank ordering of the scanned model images (frequency of response)

Machine	*Rank*	*Model* 18	19	20	21	22(resolution)
E	1st	2	19	—	—	—
	2nd	13	3	2	—	4
	3rd	5	—	11	—	6
	4th	2	—	8	1	11
	5th	—	—	1	21	1
F	1st	17	3	1	1	—
	2nd	3	18	1	—	—
	3rd	—	1	6	3	12
	4th	1	—	7	4	10
	5th	1	—	7	14	—
G	1st	3	18	1	—	—
	2nd	18	3	—	1	—
	3rd	1	1	15	5	—
	4th	—	—	5	16	1
	5th	—	—	1	—	21
H	1st	9	5	1	1	6
	2nd	9	1	2	—	9
	3rd	4	8	1	6	4
	4th	—	8	1	10	3
	5th	—	—	17	5	—

emphatically. For scanners *E* and *G* model 18 seems most similar to the liver, while for scanners *F* and *H*, model 19 seems most similar. (The fifth model image was obtained by optimizing the scan of model 18 for resolution of the wires included in the models. It thus represents a fairly haphazard setting of the scanner controls as far as texture display is concerned.)

The second part of the study, being more arduous, could only be performed with a much smaller group of unskilled volunteers between the ages of 22 and 25, consisting of three males and two females. For each subject one of the sixteen model scans was selected at random and placed in front of the subject. The remaining fifteen images were shuffled and placed one at a time to the right of the first image. Subjects were then asked to rate the similarity of the two stimuli on a seven point scale where a score of 1 indicated extreme dissimilarity and a score of 7 indicated great similarity. The whole process was repeated using each of the sixteen images as the initial stimulus, giving a total of 240 similarity judgements for each observer. The full data matrix was subjected to a non-metric multidimensional scaling analysis (114) and a two-dimensional solution of the results is shown in figure 2.11.

The most obvious feature of the data is that for scanners *E*, *F* and *H* with models 18, 19 and 20 the data form relatively distinct and well-spaced clusters. This indicates marked perceived differences between the images from the

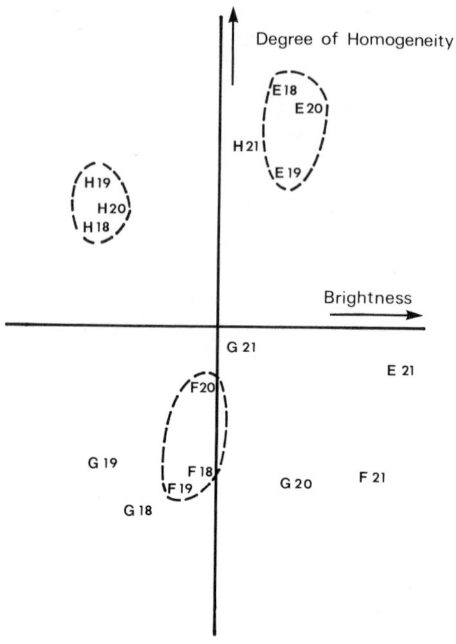

Figure 2.11 Two-dimensional solution of the multidimensional scaling analysis of image pair comparisons.

different scanners. This is an interesting result in that it confirms the qualitative conclusions of the quantitative study reported in section 2.5, as well as the theoretical discussions of section 2.3, on the prime influence of the scanner in the images obtained. This result is well known to practitioners of ultrasonic scanning, but has hitherto been anecdotal in its substance. A number of analyses on subjects of the data obtained (113) confirm the conclusions of the multidimensional scaling analysis. The axes in the display of figure 2.11 represent criteria used by the subjects for the basis of visual similarity or otherwise. Tentatively these may be labelled as a brightness dimension, and a homogeneity of structure dimension. However before these terms can be safely used they need evaluation with another independent study.

2.7 Conclusion

The brief summary presented above of the scientific basis of ultrasonic textural imaging indicates a tantalizing combination of well developed theoretical principles for the basis of the imaging process and an absence of rigorous experimental justification. The circumstantial evidence from a number of incidental observations tends to confirm the validity of the speckle analysis approach to the problem. However the assumptions of this analysis are in conflict with the preliminary reports of measurements of wave propagation in tissues. Considerable work in the measurement area is required to resolve this dilemma which urgently needs the conception and execution of critical experiments.

Even if the mechanisms of image formation are the subject of a rigorous description in the near future, the effective use of tissue models in scanner assessment will require some objective assessment of the images produced. Some of the hazards of image quantification have been exemplified in the scanner survey reported in section 2.5. The possibility of using quantitative indices to optimize the images presented for visual discrimination, irrespective of our scientific understanding of the processes involved in the image formation, is an exciting possibility. However with the continually increasing objective evidence for the dominant role of the scanner characteristics in the quantitative and visual characteristics of the images, any developments in this direction are likely to have a validity only in the context of the scanner considered. This will militate against the transfer of tissue characterization techniques based on image analysis from one clinic to another. The only way of assessing the significance of some of these potentially severe problems is the continued careful examination of the different areas in which they occur.

Acknowledgements

The author wishes to express his sincere thanks to the many colleagues who have contributed to the work described in this chapter: to Dr R. A. Bacon for making the models, performing the quantitative analysis of the images, and assistance with the computing; to Henry J. Dudley of the Department of Microbiology for the use of the image analyser; to J. A. Evans of the C & A Hospital, Bangor, for assistance in obtaining the scans for the visual analysis; to Ian Davies of the Department of Psychology for collaboration on the visual analysis; to Paul Filmore, Les Anson and Joe Aindow for assistance in taking the photographs of the model images on the scanner survey; to the manufacturers and clinicians who kindly loaned us their machines; and to Mrs Sheila Rudman for the preparation of many of the figures. Much of the work reported was funded by the D.H.S.S. whose support is gratefully acknowledged.

References

1. D. N. White, G. Clark, J. Carson and White (eds.), *Ultrasound in Biomedicine*, Pergamon Press (1982) 4.
2. C. R. Hill, *Brit. J. Radiol.* **46** (1973) 889.
3. K. T. Dussik, F. Dussik and L. Wyt, *Wien Med. Wochsr.* **97** (1947) 425.
4. J. J. Wild and J. M. Reid, *Am. J. Path.* **28** (1952) 839.
5. C. R. Hill, *J. Phys. E.* **9** (1976) 153.
6. P. Greguss, *Ultrasonic Imaging*, Focal Press, London (1980).
7. P. N. T. Wells, *Biomedical Ultrasonics*, Academic Press (1977).
8. S. A. Goss, R. L. Johnston and F. Dunn, *J. Acoust. Soc. Am.* **64** (1978) 423.
9. R. C. Chivers and R. J. Parry, *J. Acoust. Soc. Am.* **63** (1978) 940.

10. G. D. Ludwig and L. W. Struthers, *Naval Med. Res. Inst. Project NM* 004001 *Report No.* 4 (June 1949).
11. G. D. Ludwig, *J. Acoust. Soc. Am.* **22** (1950) 862.
12. A. S. Ahuja, *Ultrasonic Imaging* **1** (1979) 346.
13. A. S. Ahuja, *Ultrasonic Imaging* **1** (1979) 136.
14. R. C. Chivers, *J. Med. Eng. Tech.* **5** (1981) 128.
15. T. G. Brown, in *Proc. 2nd Int. Conf. Med. Elect.* (1960) 358 (see 7).
16. G. Baum, *J. Acoust. Soc. Am.* **48** (1970) 1407.
17. R. C. Chivers, in *Recent Advances in Ultrasound in Biomedicine* (ed. D. N. White), Research Studies Press (1977) 217.
18. N. Bom, C. T. Lancee, J. Honkoop and P. G. Hugenholtz, *Biomed. Eng.* **6** (1971) 500.
19. G. S. Kino and C. S. DeSilste, *Ultrasonic Imaging* **1** (1979) 189.
20. D. H. Howry, *Radiol. Clin. N. Am.* **3** (1965) 433.
21. R. C. Chivers, Ph.D. Thesis, University of London (1973).
22. G. Kossof, *Proc. Roy. Soc. Med.* **67** (1974) 135.
23. R. C. Chivers, *Ultrasound Med. Biol.* **7** (1981) 1.
24. R. C. Chivers and C. R. Hill, *Phys. Med. Biol.* **20** (1975) 799.
25. R. L. Smith, *Ultrasonics* **20** (1982) 211.
26. J. C. Gore and S. Leeman, *Phys. Med. Biol.* **22** (1977) 431.
27. R. C. Chivers and J. D. Aindow, *Ultrasound Med. Biol.* **8**, Supplement No. 1 (1982) 32.
28. R. C. Chivers, L. Bossellaar and P. R. Filmore, *J. Acoust. Soc. Am.* **68** (1978) 80.
29. J. D. Aindow and R. C. Chivers, *J. Phys. E* **15** (1982) 83.
30. J. C. Lockwood and J. G. Willette, *J. Acoust. Soc. Am.* **53** (1973) 735.
31. D. A. Carpenter, D. E. Robinson and G. Kossoff, in *Recent Advances in Ultrasound Diagnosis* Vol. 2, (ed. A. Kurjak) Excerpta Medica (1980) 67.
32. D. Nicholas, in *Medical Ultrasonic Images* (ed. C. R. Hill and A. Kratochwil) Excerpta Medica (1981) 61.
33. C. R. Hill and D. A. Carpenter, *Brit. J. Radio.* **49** (1976) 238.
34. P. N. T. Wells, in *Ultrasonics in Medicine* (ed. M. de Vlieger *et al.*) Excerpta Medica (1974) 30.
35. R. H. T. Bates, R. M. Lewitt, M. J. McDonnell, M. O. Milner and T. M. Peters, *Phys. Tech.* **9** (1978) 101.
36. A. Goldstein, *Radiology* **121** (1976) 157.
37. G. Rettenmaier, in *Ultrasonics in Medicine* (ed. M. de Vlieger *et al.*) Excerpta Medica (1974) 199.
38. D. C. Almond and W. I. J. Pryce, *Ultrasound Med. Biol.* **8** (1982) 29.
39. D. Nicholas, in *Recent Advances in Ultrasound Diagnosis*, Vol. 2, (ed. A. Kurjak) Excerpta Media (1980) 41.
40. J. M. Reid, in *Ultrasonic Tissue Characterization* (ed. M. Linzer) N. B. S. Special Publication 453, Washington (1976) 29.
41. S. A. Goss, R. L. Johnston and F. Dunn, *J. Acoust . Soc. Am.* **68** (1980) 93.
42. M. B. Ghitis and A. S. Khimunin, *Sov. Phys. Acoust.* **14** (1969) 413.
43. R. A. Bacon and R. C. Chivers, *Acoust. Lett.* **5** (1981) 22.
44. R. J. Parry and R. C. Chivers, in *Ultrasonic Tissue Characterization* Vol. 2 (ed. M. Linzer) N.B.S. Special Publication 525, Washington (1979) 343.
45. M. O'Donnell, E. T. Jaynes and J. G. Miller, *J. Acoust. Soc. Am.* **63** (1978) 1935.
46. M. O'Donnell, E. T. Jaynes and J. G. Miller, *J. Acoust. Soc. Am.* **69** (1981) 696.
47. S. Leeman, *Phys. Med. Biol.* **25** (1980) 481.
48. K. V. Gurumurthy and R. M. Arthur, *Ultrasonic Imaging* **4** (1982) 355.
49. W. J. Fry, *J. Acoust. Soc. Am.* **24** (1952) 212.
50. C. R. Hill, *Brit. J. Radiol.* **41** (1968) 561.
51. P. D. Southgate, *J. Acoust. Soc. Am.* **34** (1965) 480.
52. P. W. Marcus and E. L. Carstensen, *J. Acoust. Soc. Am.* **58** (1975) 1334.
53. J. K. Zieniuk and R. C. Chivers, *Ultrasonics* **14** (1976) 161.
54. L. J. Busse, J. G. Miller, D. E. Yuhas, J. W. Mimbs, A.N. Weiss snd B. E. Sobel, in *Ultrasonics in Medicine* Vol. 3 B (ed. D. N. White), Plenum Press (1977) 1519.

55. R. C. Chivers, *J. Phys.* **D13** (1980) 1997.
56. R. C. Chivers, *Ultrasonics Med. Biol.* **3** (1977) 1.
57. P. R. Filmore, J. D. Aindow and R. C. Chivers, in *Ultrasonics International* 81, IPC Press, Guildford (1981) 110.
58. J. D. Aindow and R. C. Chivers, *Acoust Lett.* **5** (1982) 144.
59. J. D. Aindow and R. C. Chivers, *J. Phys. E* **15** (1982) 1027.
60. J. D. Aindow and R. C. Chivers, *J. Phys. E* **15** (1982) 83.
61. J. D. Aindow and R. C. Chivers, *J. Acoust. Soc. Am.* (1983), in press.
62. J. D. Aindow and R. C. Chivers, *Ultrasound Med. Biol.* **8** Supplement No. 1 (1982) 1.
63. J. C. Bamber, Ph.D. Thesis, University of London (1980).
64. S. A. Goss, L. A. Frizzell and F. Dunn, *Ultrasound Med. Biol.* **5** (1979) 181.
65. R. C. Waag, and R. M. Lerner, in *Proc. IEEE Ultrasonics Symposium*, Monterey, California, Cat. No. 73 CH0807-8SU (1973) 63.
66. D. Nicholas, in *Recent Advances in Ultrasound in Biomedicine* (ed. D. N. White) Research Studies Press (1977) 1.
67. J. P. Jones, S. Leeman and J. Blackledge, in *Proc. 1st Int. Symp. Medical Imaging and Image Interpretation*, Berlin, IEEE Cat. No. 82 CH1804-4 (1982) 325.
68. R. A. Sigelmann and J. M. Reid, *J. Acoust. Soc. Am.* **53** (1973) 1351.
69. D. Nicholas, C. R. Hill and D. K. Nassiri, *Ultrasound Med. Biol.* **8** (1982) 7.
70. J. W. Mimbs, M. O'Donnell, D. Bauwens, J. G. Miller and B. E. Sobel, *Circulation Res.* **47** (1980) 49.
71. M. O'Donnell and J. G. Miller, *J. App. Phys.* **52** (1981) 1056.
72. K. K. Shung and J. M. Drzerzianowski, *Ultrasonic Imaging* **4** (1982) 56.
73. K. K. Shung, R. A. Sigelmann and J. M. Reid, *Applied Radiology* (Jan. 1976) 77, 142.
74. K. K. Shung, R. A. Sigelmann and J. M. Reid, *IEEE Trans. Biomed. Eng. BME* **23** (1977) 460.
75. D. Nicholas, in *Recent Advances in Ultrasound in Biomedicine* (ed. D. N. White) Research Studies Press (1977) 29.
76. D. Nicholas, *Ultrasound Med. Biol.* **8** (1982) 17.
77. J. M. Reid and K. K. Shung, in *Ultrasonic Tissue Characterization* Vol. 2 (ed. M. Linzer), N. B. S. Special Publication 525, Washington (1979) 153.
78. J. C. Bamber and C. R. Hill, *Ultrasound Med. Biol.* **7** (1981) 121.
79. J. C. Bamber, C. R. Hill and J. A. King, Ultrasound Med. Biol. **7** (1981) 135.
80. S. I. Fields and F. Dunn, *J. Acoust Soc. Am.* **54** (1973) 809.
81. R. Hinton and M. Dobrota, *Density Gradient Centrifugation*, North Holland Publishing Co. (1976).
82. P. Atkinson and M. V. Berry, *J. Phys. A* **7** (1974) 1293.
83. J. C. Gore and S. Leeman, Phys. Med. Biol. **22** (1977) 317.
84. C. B. Burckhardt, *IEEE Trans. Sonics and Ultrasonics* **SU25** (1978) 1.
85. J. G. Abbott and F. L. Thurstone, *Ultrasonic Imaging* **1** (1979) 303.
86. M. Fatemi and A. C. Kak, *Ultrasonic Imaging*, **2** (1980) 1.
87. J. C. Bamber and R. J. Dickinson, *Phys. Med. Biol.* **25** (1980) 463.
88. J. C. Gore and S. Leeman, *Phys. Med. Biol.* **22** (1977) 431.
89. J. P. Jones and C. Kimme-Smith, in *Acoustical Imaging* Vol. 10 (ed. P. Alais and A. F. Metherell) Plenum Press (1982) 295.
90. P. N. T. Wells and M. Halliwell, *Ultrasonics* **19** (1981) 225.
91. R. C. Chivers, in *Proc. UBIOMED IV* (ed. P. Greguss) Visegrad, Hungary (1979) 3.
92. J. Ophir and P. Jaeger, *Ultrasonic Imaging* **4** (1982) 163.
93. R. C. Chivers, *Ultrasound Med. Biol.* **2** (1976) 335.
94. K. C. Ossoinig, *J. Clin. Ultrasound* **2** (1974) 33.
95. P. Till and K. C. Ossoinig, in *Ultrasound in Medicine* Vol. 3B (ed. D. N. White) Plenum Press (1977) 2167.
96. F. G. Sommer, R. A. Filly, P. D. Edmonds, Z. Reyes and M. E. Comas, *Ultrasound Med. Biol.* **6** (1980) 135.
97. E. L. Madsen, J. A. Zagzebski, R. A. Banjavic and R. E. Jutila, *Med. Phys.* **5** (1978) 391.
98. I. S. Kol'tsova and I. G. Mikhailov, *Sov. Phys. Acoust.* **15** (1970) 390.
99. F. E. Fyke, J. F. Greenleaf and S. A. Johnston, *Ultrasound Med. Biol.* **5** (1979) 87.

100. P. D. Edmonds, Z. Reyes, D. B. Parkinson, R. A. Filly and H. Busey, in *Ultrasonic Tissue Characterization* Vol. 2 (ed. M. Linzer) N. B. S. Special Publication 525, Washington (1979) 323.
101. R. C. Eggleton and J. A. Whitcomb, *ibid.*, 327.
102. M. M. Burlew, E. L. Madsen, J. A. Zagzebski, R. A. Banjavic and S. W. Sum, *Radiology* **134** (1980) 517.
103. R. A. Lerski, T. C. Duggan and J. Christie, *Bri. J. Radiol.* **55** (1982) 156.
104. E. L. Madsen, J. A. Zagzebski and G. R. Frank, *Ultrasound Med. Biol.* **8** (1982) 277.
105. K. McCarty and W. Stewart, *Ultrasound Med. Biol.* **8** (1982) 393.
106. K. McCarty and W. Stewart, in *Proc. 3rd World Fed. Ultrasound Med. Biol.*, Brighton, (1983) in press.
107. E. L. Madsen, Z. A. Zagzebski, M. F. Insana, T. M. Burke and G. R. Frank, *Med. Phys.* **9** (1982) 703.
108. R. C. Chivers, *Arch. Acoust.* (*Warsaw*) **6** (1981) 147.
109. R. A. Bacon, R. C. Chivers, R. J. Parry and J. D. Aindow, *Ultrasonics Report* 8009, University of Surrey (1980).
110. R. C. Chivers, *Ultrasound Med. Biol.* **7** (1981) 91.
111. R. C. Chivers and R. A. Bacon, *Ultrasonics Report* 8032, University of Surrey (1980).
112. J. A. Evans and R. C. Chivers, *Ultrasonics Report* 8101, University of Surrey (1981).
113. F. M. Duarte, B.Sc. Dissertation, University of Surrey (1981).
114. R. N. Shepard, *Multidimensional Scaling : Theory and Applications in the Behavioural Sciences*, Vols. 1 and 2, (1972).

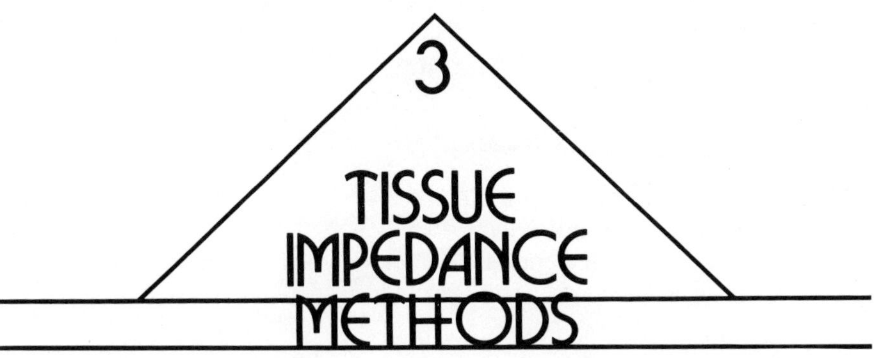

Brian H. Brown

3.1 Electrical properties of tissue

Biological tissue contains free charge carriers so that it is meaningful to consider it as an electrical conductor and to describe it in terms of a conductivity. Conductivity is a spatial variable so that it is reasonable to consider the possibility of producing images of the distribution of conductivity within the body.

Bound charges are also present in tissue so that dielectric properties also exist and can be expected to give rise to displacement currents when an electric field is applied. These properties might arise as electronic or nuclear polarization in non-polar material, as a relative displacement of negative and positive ions when these are present, or as a result of a molecular electric dipole moment where there is a distribution of positive and negative charge within a molecule. These effects may be described in terms of a relative permittivity (dielectric constant) which will be a spatial variable within the body. It is therefore reasonable to consider the possibility of producing images of the distribution of relative permittivity within the body.

In addition to the above two passive electric properties, biological tissue contains mechanisms for the active transport of ions. This is an important mechanism in neural function and also in membrane absorption processes, such as those which occur in the gastrointestinal tract.

This chapter will concern itself almost exclusively with the measurement of conductivity, which is the dominant factor when relatively low frequency (less than 100 kHz) electric fields are applied to tissue. An excellent coverage of dielectric effects is given by Pethig (1). Whilst the idea of producing spatial images of conductivity is attractive it is extremely difficult; no practical systems have appeared in the literature. What follows contains background information and outlines the problems involved in imaging and concludes with a description of a method of image reconstruction which avoids some of the problems and is able to produce images.

3.1.1 *Frequency-dependent effects*

The electrical properties of a material can be characterized by an electrical conductivity σ and permittivity ε. If a potential V is applied between the opposite faces of a unit cube of the material then a conduction current I_c and displacement current I_D will flow, where

$$I_c = V\sigma; \quad I_D = \frac{dV}{dt}\varepsilon\varepsilon_0$$

and where ε_0 is the dielectric permittivity of free space with the value $8.854 \times 10^{-12}\,\mathrm{F.m^{-1}}$. If V is sinusoidally varying then I_D is given by:

$$I_D = V \cdot 2\pi f \varepsilon\varepsilon_0$$

where f is the frequency of the sinusoidal potential.

Both conductivity and permittivity vary widely between different biological tissues (see section 3.1.4) but figure 3.1 shows a typical frequency variation of I_c and I_D for soft biological tissue. I_c increases only slowly with increasing frequency and indeed at frequencies up to 100 kHz conductivity is almost constant. I_D increases much more rapidly with increasing frequency and above about 10^8 Hz the displacement current exceeds the conduction current. Permittivity decreases with increasing frequency and there are in general three regions where rapid changes take place. The region around 10 Hz is generally considered to arise from dielectric dispersion associated with tissue interfaces such as membranes; the region around 1 MHz is

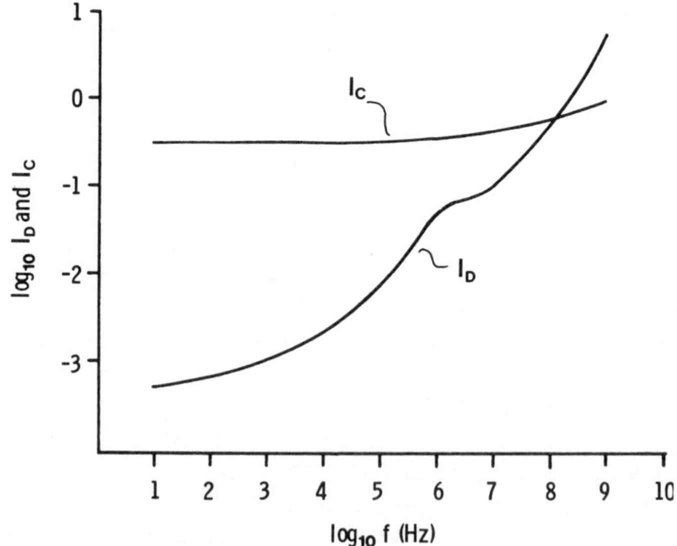

Figure 3.1 The changes in conduction current (I_c) and displacement current (I_D) with frequency for a typical biological tissue.

associated with the capacitance of cell membranes; the region around 10^{10} Hz represents the dielectric dispersion associated with polarizability of water molecules in tissue.

Inspection of fig. 3.1 shows that for relatively low frequencies (less than 100 kHz) the displacement current is likely to be very much less than the conduction current and it is therefore reasonable to neglect dielectric effects in treating tissue impedance.

3.1.2 *Biological effects*

If low-frequency currents are passed between a pair of electrodes placed on the skin then a current can be found at which sensation occurs. In general this threshold of sensation rises with increasing frequency of applied current, as shown in figure 3.2. Three fairly distinct types of sensation occur as frequency increases.

At very low frequencies (below 0.1 Hz) individual cycles can be discerned and a 'stinging sensation' occurs underneath the electrodes. The major effect is thought to be electrolysis at the electrode/tissue interface where small ulcers can form with currents as low as 100 μA. The application of low frequency currents can certainly cause ion migration and this is the mechanism of iontophoresis. Current densities within the range 0–10 Am^{-2} have been used to administer local anaesthetics through the skin, and also therapeutic drugs for some skin disorders. The applied potential acts as a forcing function

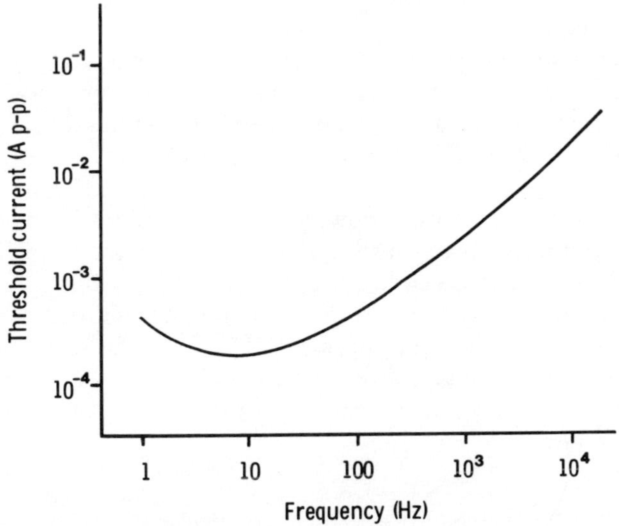

Figure 3.2 Threshold of sensation as a function of frequency for an electric current applied between 5 mm wide band electrodes encircling the base of adjacent fingers. (Result from one normal subject).

but will not cause drugs to penetrate skin which is normally impervious. The drug must normally be soluble in both fat and water; sweat ducts are the principal path for ion movement.

At frequencies above 10 Hz, electrolysis effects appear to be reversible and the dominant biological effect is that of neural stimulation. If the electrodes are placed over a large nerve trunk such as the ulnar or median, then the first sensation arises from the most rapidly conducting sensory fibres. If the amplitude of the current is increased, then more slowly conducting fibres are stimulated and motor contractions occur. Stimulation over a nerve trunk arises as a result of depolarization at a node of Ranvier. The capacitance of a single node is of the order 10 pF such that a charge of 10^{-12} coulomb is required to remove the normally occurring polarization potential of about 0.1V. 10^{-12}C can be delivered as a current of 10^{-9}A for 1 ms. However, when the current is delivered through relatively distant surface electrodes only a very small fraction of the current will pass into a particular node of Ranvier. It is therefore to be expected that the threshold shown in figure 3.2 will fall as the electrodes are moved closer to a nerve trunk.

Propagation of a nerve action potential is controlled by the external currents which flow around the area of depolarization and so depolarize adjacent areas. This depolarization of adjacent areas is not instantaneous as time is required to remove the charge associated with the capacitance of the nerve membrane. If the frequency of an applied external field is such that an action potential cannot be propagated within one cycle then neural stimulation will not occur. It is for this reason that the threshold of sensation rises as the frequency of the applied current is increased.

At frequencies above about 10 kHz the current necessary to cause neural stimulation is such that heating of the tissue is the more important biological effect. Displacement currents are usually negligible within the range 10 kHz–100 kHz and therefore the I^2R losses are dominant.

The major biological effects within our frequency range of interest are therefore electrolysis, neural stimulation and heating. In figure 3.2 the threshold sensation is given in terms of the total current which is passed between a pair of surface electrodes. The threshold will depend upon the electrode area as there is ample evidence to show that current density rather than current is the important parameter. However, the relative magnitude of the three effects we have considered is not changed when current density rather than current is used. A typical value of current density at threshold and 50 Hz is 2 Am^{-2}.

3.1.3 *Evidence for and against assumption of linearity*

If we are to characterize tissue in terms of permittivity and conductivity then it is important to be able to assume linearity, i.e. permittivity and conductivity to be independent of electric field.

Pethig (1) states that most dielectrics are linear for field strengths up to about $10^5\,\mathrm{Vm^{-1}}$. However, the field strengths across many biological membranes may be $10^6\,\mathrm{Vm^{-1}}$ and there is ample evidence to show that the properties of neural membranes are non-linear. But the membrane impedance may well not be the dominant factor when bulk measurements of tissue impedance are made. It was stated earlier that the contribution to displacement current from neural membranes does not become important until frequencies of about 1 MHz are applied.

Tissue impedance measurements made at frequencies below 100 kHz and at current densities less than $10\,\mathrm{mA\,cm^{-2}}$ ($100\,\mathrm{Am^{-2}}$) show no evidence for any significant non-linearity of conductivity. Using a four-electrode system (see section 3.3.1) applied to the human arm, it can be shown that within the frequency range 100 Hz to 100 kHz and current density range $1\,\mathrm{Am^{-2}}$ to $100\,\mathrm{Am^{-2}}$, tissue conductivity is constant to within $\pm 5\%$.

If non-linearities are present, then if a sinusoidal current is applied the resultant voltage will contain both the fundamental frequency of the applied sine wave and also harmonics. Currents applied to the human arm at current densities up to $100\,\mathrm{Am^{-2}}$ show no significant harmonic distortion (<50 dB below the fundamental) within the frequency range 100 Hz to 1 kHz.

Witsoe and Kinnen (2) claim to show linearity of human lung tissue at current densities of $800\,\mathrm{Am^{-2}}$ at 100 kHz. Geddes, using defibrillation discharges across the thorax, shows a small reduction in resistivity at current densities of $16\,000\,\mathrm{Am^{-2}}$.

Table 3.1 Typical values for the resistivity of a range of biological materials at body temperature. The frequency at which the measurements were taken is also given.

Tissue	*Resistivity*	*Frequency*
CSF	0.650 Ωm	1 kHz–30 kHz
Blood	1.48–1.76 Ωm	1 kHz–100 kHz
Skeletal muscle	1.25–3.45 Ωm (longitudinal) 6.75–18.0 Ωm (transverse)	100 Hz–1 kHz
Lung	12.75 Ωm (average of inspiration and expiration) 8–17 Ωm (expiration and inspiration)	100 kHz
Neural tissue	5.8 Ωm (brain)	—
	2.8 Ωm (grey matter)	—
	6.8 Ωm (white matter)	—
Fat	20 Ωm	1 kHz–100 kHz
Bone	> 40 Ωm	—

3.1.4 *Resistivity of various biological tissues*

There is a large literature on the electrical resistance of biological material. An excellent, although now rather old, review is given by Geddes and Baker (3). There are many discrepancies in the reported values but this is not surprising in view of the great difficulties in making measurements *in vivo* and the problems of preserving tissue for measurement *in vitro*. Table 3.1 gives typical values for a range of biological materials at body temperature (resistivity can be expected to fall by 1–2% per °C).

3.1.5 *Effect of geometry and anisotropy*

The resistance R between the faces of a cylinder of length l and cross sectional area a is given by:

$$R = \frac{\rho l}{a} \tag{3.1}$$

where ρ is the resistivity of the medium. If the system changes in either length or area then:

$$\Delta R = \left(\frac{\partial R}{\partial l}\right)_a \cdot \Delta l + \left(\frac{\partial R}{\partial a}\right)_l \cdot \Delta a \tag{3.2}$$

If the change is only in length then:

$$\frac{\Delta R}{\Delta l} = \left(\frac{\partial R}{\partial l}\right)_a = \frac{\rho}{a}$$

$$\frac{\Delta R}{\Delta V} = \frac{R}{V} \tag{3.3}$$

where V is the volume la.

If the change is in area then

$$\frac{\Delta R}{\Delta a} = \left(\frac{\partial R}{\partial a}\right)_l = -\frac{\rho l}{a^2}$$

$$\frac{\Delta R}{\Delta V} = -\frac{R}{V} \tag{3.4}$$

If the cylinder expands equally both radially and longitudinally then it is relatively easy to show that:

$$\frac{\Delta R}{\Delta V} = -\frac{R}{3V} \tag{3.5}$$

Equations (3.3), (3.4) and (3.5) can be used as the basis for interpretation of impedance plethysmography measurements. As arterial blood enters a limb during systole the limb will expand, whereas during diastole the arterial flow may be close to zero and so the limb contracts due to venous run-off. However, these equations show that the changes which can be expected to occur in resistance of the limb as the limb volume changes depend upon the manner in which the limb expands. The assumption is usually made that the limb expands radially, in which case equation (3.4) allows the changes in limb volume to be determined. Nonetheless there is no convincing evidence to rule out the possibility of longitudinal expansion of soft tissues.

Changes in limb resistance can also occur if the resistivity of the tissues changes. As blood has a resistivity about one-third that of the average resistivity of a limb, it might be expected that this will also cause resistance to fall during systole when the volume of blood in the limb is increasing.

Many tissues contain well-defined long fibres, skeletal muscle being the best example, so that it might also be expected that conductivity would be different in the longitudinal and transverse directions. This is indeed the case, and Burger and van Dongen (4) have shown that the transverse resistivity may be ten times greater than the longitudinal resistivity. The effect of anisotropy will be further considered in 3.3.3.

3.2. Measurements *in vitro*

3.2.1 *Electric equivalent circuits*

Nearly all the methods used to measure the impedance of biological tissue involve making an electrode contact to the tissue. The electrode forms a transducer between the ionic flow in the tissue and the electronic flow within the recording equipment. This transducer will have a finite impedance which cannot be neglected when making tissue resistivity measurements and it is therefore necessary to consider the electrical equivalent circuit of the transducer.

The electrochemical changes which occur underneath the electrode may be either reversible or irreversible, and in the latter case it can be expected that an electrode will polarize. Polarization can be expected to alter the equivalent circuit of the electrodes. Ag/AgCl electrodes give a reasonably stable contact as the major reversible reaction which takes place at the tissue interface is

$$\mathrm{Ag} + \mathrm{Cl}^- \leftrightarrows \mathrm{AgCl} + e^-$$

Chloride ions form charge carriers within the tissue.

There is an extensive literature on the equivalent circuit of electrodes (5.6)

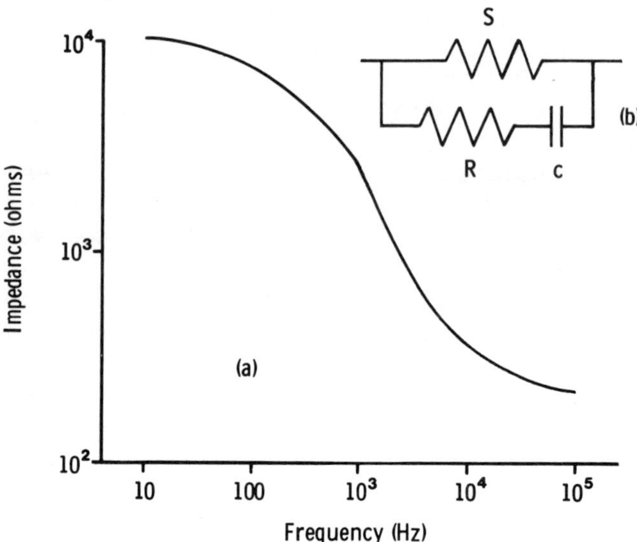

Figure 3.3 (*a*) The general form of impedance as a function of frequency for electrodes applied to skin. (*b*) A simple approximately equivalent circuit.

all of which gives an impedance which changes with frequency as shown in figure 3.3(*a*). This graph shows the impedance between a pair of electrodes placed on skin. The simplest equivalent circuit is a combination of two resistors and a capacitor as shown in figure 3.3(*b*).

At low frequencies the circuit impedance becomes S, as $1/\omega c \to \infty$ and at high frequencies becomes $SR/(S+R)$ as $1/\omega c \to 0$. Figure 3(*b*) is by no means the only possible configuration of the three components and many more components are required if a close fit with experimental results is demanded.

A pair of 1 cm diameter Ag/AgCl electrodes placed on well abraded skin over the forearm will give typical values for S, R and C: $S = 10\,\mathrm{k}\Omega$, $R = 200\,\Omega$, $C = 50\,\mathrm{nF}$. Changes in the preparation of the skin surface will affect S and C to a greater extend than R and a change in electrode area will usually give a corresponding change in C but an inverse change in S and R.

This chapter is concerned with tissue impedance and not electrode impedance. However, as almost every method of measuring tissue impedance involves an electrode it must be borne in mind that electrode impedances are usually very much larger than tissue impedances. They add greatly to the problem of making precise measurements of tissue impedance.

3.2.2 *Conductivity cells*

Conductivity cells can be used to measure the resistivity of body fluids. They consist of a small cell and then either two or four electrodes.

Equation (3.1) gave the resistance between the end faces of a cylindrical conductor as

$$R = \frac{\rho l}{a} = \frac{l}{\sigma a}$$

where ρ is the resistivity of the material, σ the conductivity, l the length and a the cross-sectional area. If R can be measured, then

$$\sigma = \frac{1}{R} \cdot \frac{l}{a} \tag{3.6}$$

In the case of a conductivity cell, R is the resistance of the cell containing the fluid and l/a is a constant depending upon the dimensions of the cell.

Equation (3.6) applies to a resistance measured between the parallel faces of a block of material. Most conductivity cells use circumferential electrodes placed at the top and bottom of a cylindrical cell. A cell constant can be determined either theoretically or experimentally for a particular cell and so the conductivity of a filling fluid calculated as

$$\sigma(\text{siemens}) = \frac{1}{(\text{cell resistance } (\Omega))} \times \text{cell constant } (\text{m}^{-1})$$

This simple two-electrode system is widely used for measurements on fluids such as blood and urine. However, errors can arise when making measurements on fluids of high conductivity where electrode impedance becomes significant. The care with which electrodes are cleaned and prepared affects

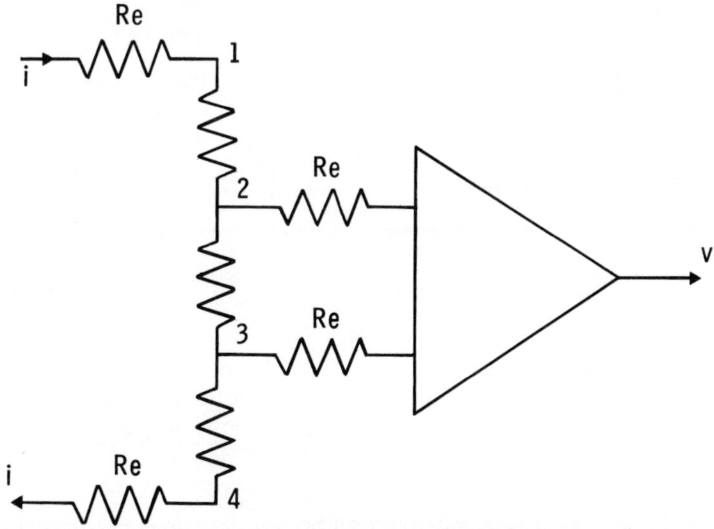

Figure 3.4 Equivalent circuit of a four-electrode conductivity cell.

the errors. A four-electrode system can considerably reduce these errors; this is shown schematically in figure 3.4. A constant current (usually within the frequency range 50 Hz to 5 kHz) is passed between the outer pair of electrodes 1 and 4. This will give rise to a potential drop across the electrode resistances R_e and the resistance of the fluid in the three sections 1–2, 2–3, 3–4 of the cell.

The potential drop V across the section 2–3 of resistance S can be measured using a high input impedance amplifier and the result will not be significantly affected by R_e. In the case of figure 3.4, S is given by V/i. In the case of the conductivity cell the conductivity of the fluid is given by

$$\sigma = \frac{i}{V} \times \text{cell constant } (\text{m}^{-1})$$

3.3. Measurements *in vivo*

3.3.1 *Four-electrode systems—homogeneous case*

Consider four electrodes on the surface of a homogeneous, isotropic, semi-infinite medium as shown in figure 3.5.

If positive and negative current sources of strength I are applied to electrodes 1 and 4 then we may determine the resulting potential V between electrodes 2 and 3. The four electrodes are equally spaced and are collinear. The current I will spread out radially from 1 and 4 such that the current density is $I/2\pi r^2$ where r is the radial distance. The current density along the line of the electrodes is given by

$$\text{current density} = \frac{I}{2\pi x^2} + \frac{I}{2\pi(3a - x)^2}. \qquad (3.7)$$

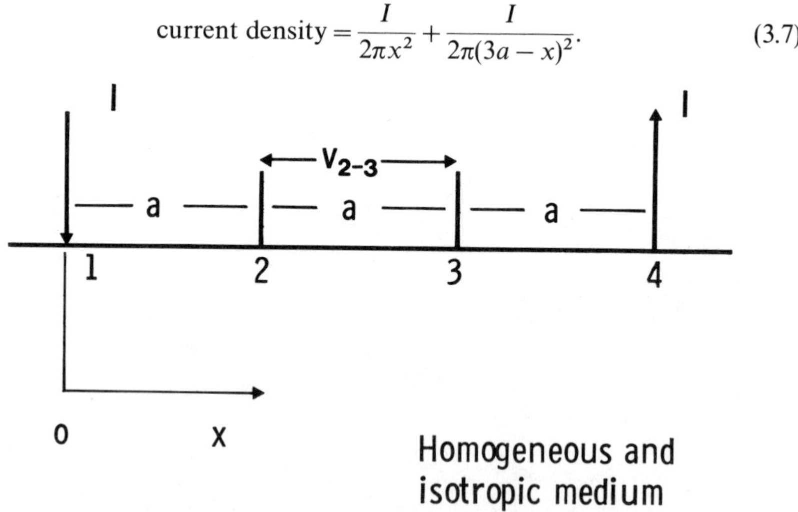

Figure 3.5 Four-electrode system used to determine resistivity ρ of a conducting medium.

If we consider an element of length Δx and cross sectional area Δa of the tissue with resistivity ρ then

$$\Delta R = \rho \frac{\Delta x}{\Delta a}$$

and the potential drop ΔV along this element is

$$\Delta V = \left[\frac{I}{2\pi x^2} + \frac{I}{2\pi(3a-x)^2} \right] \cdot \Delta a \cdot \rho \cdot \frac{\Delta x}{\Delta a}.$$

Integrating between electrodes 2 and 3 we obtain

$$V_{2-3} = \int_a^{2a} \frac{\rho I}{2\pi} \left(\frac{1}{x^2} - \frac{1}{(3a-x)^2} \right) \cdot dx$$

$$V_{2-3} = \frac{\rho I}{2\pi a}$$

and

$$\rho = \frac{2\pi a}{I} \cdot V_{2-3}. \tag{3.8}$$

The resistivity of the medium can therefore be simply obtained by measuring V_{2-3}.

The same result can be obtained by analogy with the case of an electrostatic dipole as Laplace's equation $\nabla^2 V = 0$ applies in both cases to give

$$V = \frac{\rho I}{2\pi} \left(\frac{1}{r_1} - \frac{1}{r_4} \right) \tag{3.9}$$

where r_1 and r_4 are the distances of the point of potential measurement to the current electrodes 1 and 4 respectively.

This experimental system has been widely used for measurements of resistivity ρ. However, errors can arise from high current densities at the drive electrodes and local non-homogeneities.

An equation for the potential V_p at any point on the surface of a semi-infinite medium when current is applied through a pair of circular disc electrodes has been derived by Witsoe (7). This gives

$$V_p = \frac{\rho I}{2\pi a} \left(\sin^{-1} \frac{a}{r_1} - \sin^{-1} \frac{a}{r_4} \right) \tag{3.10}$$

where a is the radius of the discs and r_1 and r_4 are the distances from P to the current drive electrodes. This equation can be used with relatively large drive electrodes to minimize the effect of high current densities under these electrodes. When $a \ll r_1$ and r_4, equation (3.10) reduces to equation (3.9).

3.3.2 *Four-electrode system—non-homogeneous case*

Treatment of the general case for a spatial distribution of resistivity is extremely difficult. However, simple geometries can offer an analytic solution.

If the semi-infinite medium of figure 3.5(*b*) is replaced by a layer of resistivity ρ_1 and depth *b*, overlying a semi-infinite medium of resistivity ρ_2, then the potential V_{2-3} can be shown (8) to be

$$V_{2-3} = \frac{\rho_1 I}{2\pi a}\left[1 + 4a \sum_{n=1}^{\infty} \beta^n \left(\frac{1}{\sqrt{[(a)^2 + (2bn)^2]}} - \frac{1}{\sqrt{[(2a)^2 + (2bn)^2]}}\right)\right] \tag{3.11}$$

where

$$\beta = \frac{\rho_2 - \rho_1}{\rho_2 + \rho_1}. \tag{3.12}$$

When ρ_1, *a*, *b* and *I* are known then ρ_2 may be determined. If both ρ_1 and ρ_2 are unknown then two measurements of V_{2-3} at different electrode spacings provide two equations from which ρ_1 and ρ_2 may be found.

Stefanesco and Schlumberger (9) extended this treatment to many layers of different resistivities. This technique has been applied to geophysical measurements of the earth's resistivity at different depths. However, surface, or near surface, inhomogeneities can add serious errors to this technique.

3.3.3 *Measurement of anisotropy*

The techniques considered in the previous two sections can be applied to electrodes placed longitudinally on a relatively large limb. However, skeletal muscle is not isotropic—the resistivity is much greater in the transverse direction than in the longitudinal direction—and we will therefore consider the problem of transverse measurements of resistivity.

Consider a limb with long electrodes placed diametrically opposite and parallel to the axis of the limb (figure 3.6). Each electrode consists of several smaller electrodes with large resistances placed in series such that the current lines are in planes at right angles to the axis of the limb.

Because the current paths are constrained to only two dimensions we can consider a homogeneous circular slab of thickness *h* which the current enters at 1 and 2 (figure 3.7) and we make measurements of the potential at points 3 and 4.

It can be shown that the potential on the circumference is given by

$$V = \frac{\rho I}{\pi h} \ln \frac{r_1}{r_2} \tag{3.13}$$

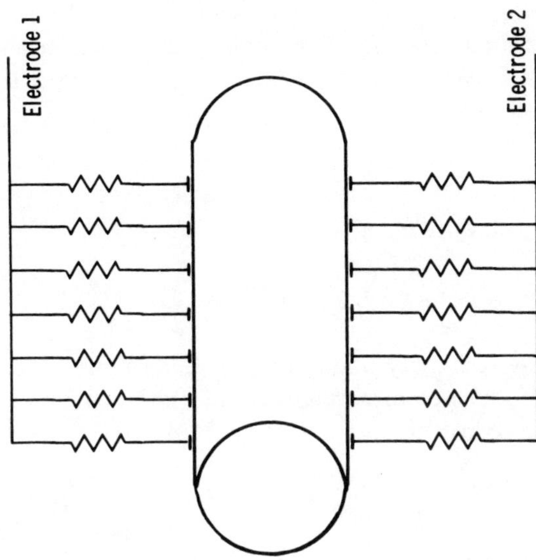

Figure 3.6 Two electrodes placed on a limb. Each electrode consists of many smaller ones such that current lines are in planes at right angles to the axis of the limb.

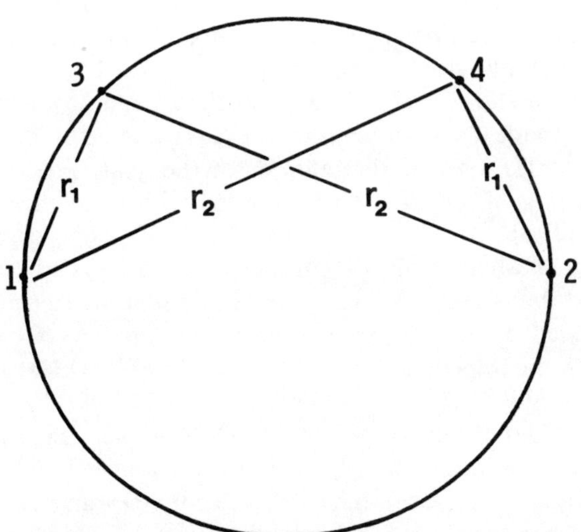

Figure 3.7 A disc of material. Current is injected between points 1 and 2 and the potential recorded between 3 and 4.

where ρ is the resistivity, I the current flowing and r_1 and r_2 the distances to the point on the circumference from the points of current injection.

The potential between 3 and 4 when these are symmetrically spaced with respect to 1 and 2 is

$$V_{3-4} = \frac{2\rho I}{\pi h} \ln \frac{r_1}{r_2}$$

and if the electrodes are equally spaced,

$$V_{3-4} = \frac{2\rho I}{\pi h} \ln \sqrt{3} \tag{3.14}$$

The transverse resistivity ρ may thus be determined. A correction can be applied for the presence of a relatively insulating bone at the centre of a limb (4) but the correction factor is very small.

3.3.4 *Impedance plethysmography*

Impedance plethysmography is a technique whereby a bulk impedance measurement is used to determine either the change in blood volume within the total tissue volume or changes in the geometry of the volume. In both cases the aim is to determine changes in blood volume and hence determine flow as the rate of change of volume. The dependence of limb impedance upon system geometry was considered briefly in section 3.1.5. Only the problem of impedance measurement will be considered here and not the determination of blood flow from these changes. Some references to the latter are given in the bibliography (10, 11).

If four circumferential electrodes are attached to a limb and a measurement of impedance made as described in the previous section, then a typical impedance is 100 Ω and the fluctuations at the heart rate are about 0.1 Ω. These fluctuations of 0.1% must be recorded with great care if artefacts are not to be significant.

In figure 3.8 an alternating current I_{p-p} at 100 kHz is applied to the outer electrodes and the potential U across the inner pair of electrodes is recorded via the amplifier A. Resistances R represent electrode contact impedances (typically 1 kΩ; the impedance is usually resistive at 100 kHz) and S the tissue resistance (typically 50 Ω).

We will consider firstly the effect of *amplifier input impedance* Z_{in}. It is relatively easy to obtain an amplifier input resistance of 10^9 ohms in which case the potential drop across the electrode contact impedance is negligible. However, between the leads connecting the electrodes to the amplifier there might well be a capacitance of 100 pF which presents an impedance of $1/\omega c = 20$ kΩ at 100 kHz, therefore the effective Z_{in} is only 20 kΩ and we can

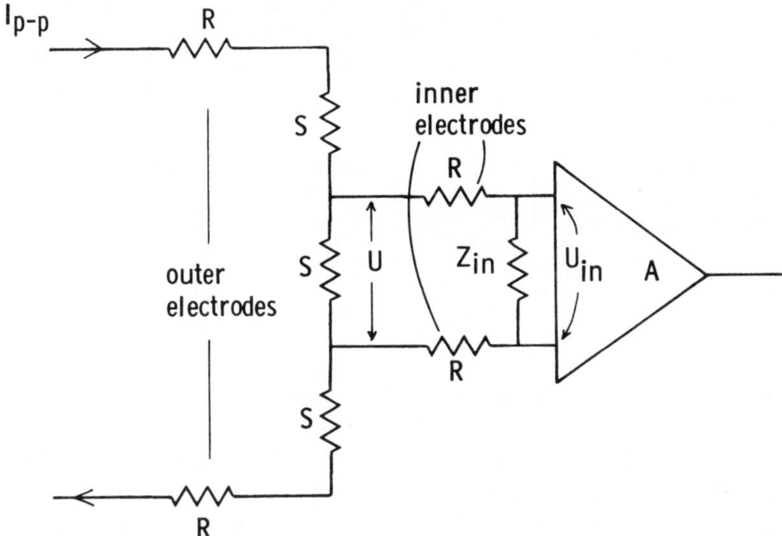

Figure 3.8 Configuration of tissue impendances S and electrode impedances R in a four-electrode impedance plethysmograph.

expect a potential drop across the electrode impedances of up to 10%. If the electrode impedances fluctuate with the heartbeat, as they probably will because the electrochemical equilibrium at the electrode interface is disturbed by movement, then U_{in} will fluctuate and these fluctuations might well be larger than the fluctuations in U which we wish to record. This is a major problem which can only be reduced by using screened input cables where the screen is driven to reduce input capacitance. With care input capacitance can be reduced to less than 5 pF.

We have only considered differential input impedance here. A similar argument can be used to show that the common mode input impedance must also be made $\gg R$.

Another important factor is the stability of the current source I_{p-p}. An instability of 0.1% in this current will give 0.1% of noise on potential U which cannot be distinguished from the changes resulting from tissue resistivity changes.

A block diagram of an impedance plethysmograph is given in figure 3.9. Comments on this diagram are as follows:

(a) The input amplifier is a bandpass type to reject low frequency bioelectric signals. The differential and common mode input impedances must be high at 100 kHz.

(b) The oscillator and current source must have an amplitude stability of at least 0.01%.

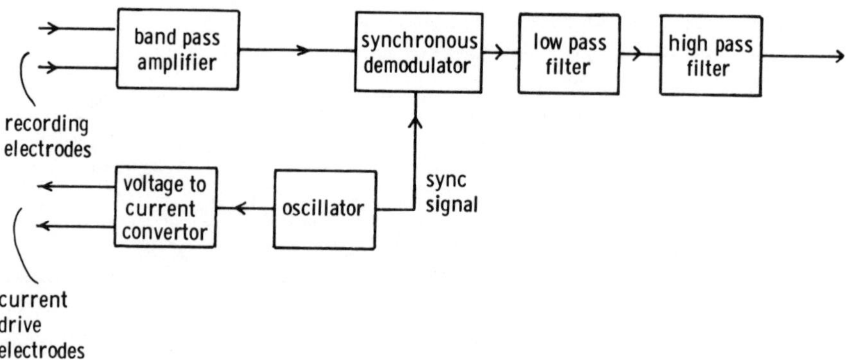

Figure 3.9 Block diagram of an impedance plethysmograph.

(c) A synchronous demodulator is much better than a rectifier as it will reject signals which are not in phase with the oscillator output.
(d) A low-pass filter removes any residual signal at the oscillator frequency.
(e) The high-pass filter removes the d.c. component of the signal to leave only the fluctuation in U resulting from blood flow changes.
An alternative to the high-pass filter is a d.c. back-off facility.

3.3.5 *Impedance pneumography*

Impedance pneumography uses the changes in thoracic impedance during respiration as a measure of respiratory function. The changes in impedance result from lung volume changes, from changes in shape of the thorax and from changes in the volume of blood within the thorax during respiration. There is no satisfactory method which allows lung volume to be determined from the impedance changes. However, thoracic impedance changes can be recorded relatively easily and give a signal which can be used to monitor respiratory rate and fluctuations in respiratory volume (12).

The changes in impedance with respiration from electrodes placed just below the axilla are typically 1 Ω on a base impedance of 50 Ω at 100 kHz. The size of the fluctuations varies with posture and both the size and shape of the subject. The equipment described in section 3.3.4 for impedance plethysmography can be used for impedance pneumography but because the respiratory signals are much larger than the cardiac changes, less care is required in the design of the amplifier. As the impedance measurements are made at a relatively high frequency it is impossible to make measurements through the same electrodes as are used for ECG recordings in routine patient monitoring.

There is a well documented change in lung impedance from inspiration to expiration and attempts have been made to use guard electrode systems to measure regional differences in thoracic impedance. However, there is considerable doubt about the accuracy of measurements made using guard electrode systems (13) particularly those made on such an unhomogeneous structure as the thorax, and this is considered further in section 3.4.1.

3.4 Spatial resolution—imaging

It is attractive to consider the possibility of determining the spatial distribution of resistivity within an object such as the human body where there is good differentiation of a wide range of tissues (see table 3.1). Such a technique would also find applications in geological exploration; there is a considerable literature on the use of the techniques described in section 3.3.2 to determination of the changes of ground resistivity with depth.

3.4.1 *Electrode guarding*

The problem to be overcome in making impedance images is that caused by the divergence of field lines (current) when a potential is applied between two points on the conductor. In X-ray CT imaging the value assigned to a point within the image during the reconstruction process (see section 3.1 of Vol. 1) is mainly determined by the line integrals which go through or near that point. These line integrals are accessible since a collimated beam of X-rays can be passed through the object. However, when an electrical current is applied to tissue the field pattern is complex and we do not have access to the necessary line integrals.

Two approaches are possible. The first is to attempt to constrain the field lines to parallel paths within the object, so that standard CT reconstruction techniques can be used. The second is to use an algebraic image reconstruction technique which can cope with complex field patterns. Electrode guarding is the way in which the first approach has been attempted.

If a potential V is applied to the resistive matrix of figure 3.10 at point B then current will flow through all the elements of the matrix and the total current flowing I will be a function of all the elements. However, if an identical 'guard potential' V is applied to points A and C and points D, F returned to zero potential then, under certain conditions, there will be no current through the R elements, and

$$\frac{V}{I} = S_{21} + S_{22} + S_{23}$$

The total resistance seen by the central electrodes is thus the line integral of resistance between the electrodes.

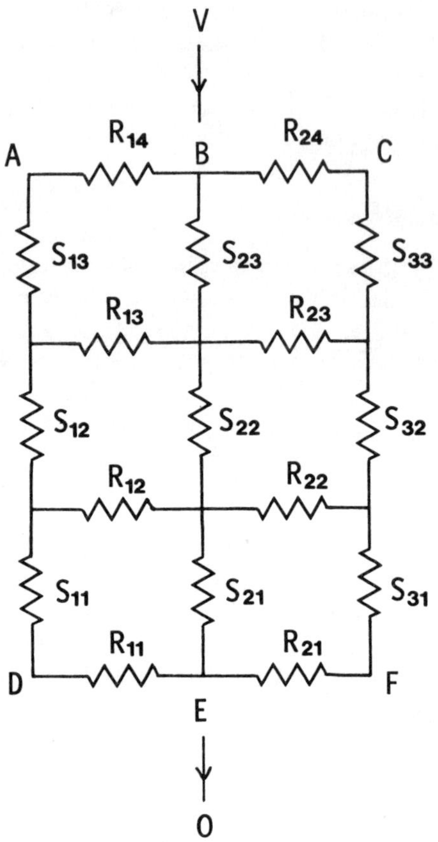

Figure 3.10 A resistive matrix with potential V applied between points B and E.

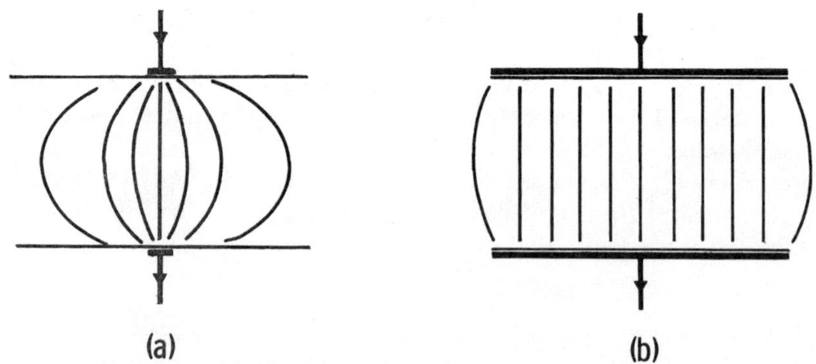

Figure 3.11 The field pattern (current) to be expected when small electrodes (*a*) and large electrodes (*b*) are applied to a slab of conductor.

The condition to be met is that the form of potential drop along the paths $A-D$ and $C-F$ is identical with that along BE such that there is no lateral current flow. This condition can be met if the matrix is homogeneous i.e. all resistance values are equal. The minimal condition is that

$$\frac{S_{am}}{\sum_{m=1}^{n} S_{am}} = \frac{S_{bm}}{\sum_{m=1}^{n} S_{bm}} \tag{3.15}$$

for all values of a and b from 1 to n and where the matrix has been extended to m rows and n columns.

Guarding can be effected by applying large electrodes to the conducting

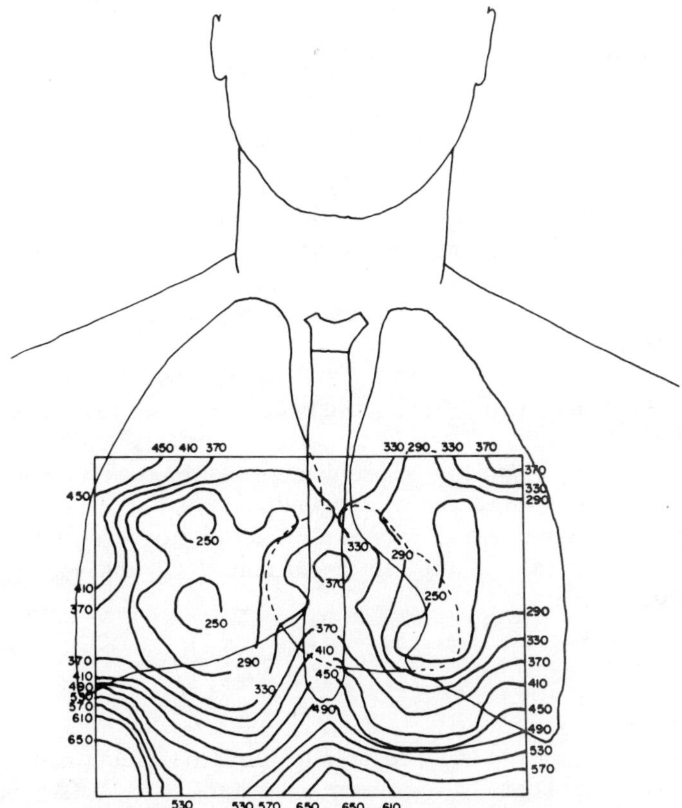

Figure 3.12 Isoadmittance contour map of human thorax. Admittance values are in units of microsiemens per electrode. An array of 144 mutually guarding electrodes was used to produce this result. From R. P. Henderson and J. G. Webster, *IEEE Transactions on Biomedical Engineering*, BME-25, No. 3 (1978).

volume. Figure 3.11 shows the field patterns to be expected when small and large electrodes are applied to a slab of conductor.

If in figure 3.11(*b*) the lower electrode is replaced by many small electrodes then by measuring the current flow through these electrodes the line integrals of resistance along the columns of conductor between the small and large electrodes can be recorded. However, it must be remembered that this is only the case if our assumption of parallel field lines is met, and this is only the case for a homogeneous medium or one where equation (3.15) is met. If a non-homogeneity is introduced then the field lines will be distorted.

Using a large electrode on one side of the thorax and a matrix of 100 electrodes on the other, Henderson and Webster (14) have produced a transmission image of thoracic impedance (figure 3.12). The resolution of this type of image is poor and Plonsey and Collin (13) show that, in addition, the distortion of field lines in a non-homogeneous structure such as the thorax will give rise to image distortion and spatial translation of structures.

3.4.2 *Image reconstruction from resistance measurements*

A number of workers have collected data from a guarded system as described in the previous sections (15, 16, 17) and attempted image reconstruction, but with very disappointing results. Schomberg (15) says that image reconstruction is very susceptible to errors in the data, and when using a system of an impressed potential and measurements of exit currents electrode impedances can cause significant errors. Schneider (18) is quoted as showing that there exist infinitely many admissible spatial distributions of resistivity which give the same peripheral currents as are obtained for a homogeneous conductor.

Bates *et al.* (19) formulate the problem in two dimensions as follows. An arbitrary point P within a circle has cylindrical polar co-ordinates (r, θ) and we consider a conductivity distribution $\sigma = \sigma(r, \theta)$. For a resistive system we can neglect temporal variations in the current $J(r, \theta)$ and voltage $V = V(r, \theta)$ and so obtain

$$J = -\sigma \nabla V \quad \text{and} \quad \nabla J = 0 \tag{3.16}$$

or

$$\nabla^2 V + \nabla V \Delta \tau = 0 \tag{3.17}$$

where $\tau(r, \theta) = \ln(\sigma)$. Eq. (3.17) reduces to Laplace's equation $\nabla^2 V = 0$ in a homogeneous medium. The problem of imaging is to determine τ from measurement of $J(r, \theta)$ given the voltage $V(r, \theta)$ applied to the circumference. Bates *et al.* go on to consider the situation where $V(1, \theta) = \cos\theta$ which forces the current field lines to be parallel to the line joining (1, 0) and (1, π). This is the situation when perfect electrode guarding is applied to a homogeneous medium.

If a system is now considered where the conductivity is σ_0 from the centre to a radius α and then 1 to the circumference it is shown that $J(1,\theta)$ is a function of σ_0, α^2 and θ. Because $J(1,\theta)$ is the only quantity available for measurement then σ_0 and α cannot be determined without *a priori* information. Under the conditions considered, therefore, there is a basic limitation on the method for determination of $\sigma(r,\theta)$.

However, Bates *et al.* go on to consider the situation where more extensive sets of measurements are made by varying $V(1,\theta)$ and conclude that a unique solution for $\sigma(r,\theta)$ may then be possible.

3.4.3 *Image reconstruction from potential distribution measurements*

The major conclusion of the foregoing two sections is that guard electrode systems do not offer an easy route to impedance imaging. We will now consider a system based upon the recording of peripheral potentials, when a current is imposed on the conductor, and subsequent back projection of these potentials along curved isopotentials. For simplicity we will consider only two dimensions at this stage.

Figure 3.13 shows the form of the isopotentials to be expected when current I is imposed on a homogeneous disc of material. The potential distribution may be obtained from Laplace's equation, which is given in cylindrical polar co-ordinates by:

$$r\frac{\partial}{\partial r}\left(r\cdot\frac{\partial V}{\partial r}\right)+\frac{\partial^2 V}{\partial\theta^2}+\frac{\partial^2 V}{\partial z^2}=0$$

In our two-dimensional case

$$\frac{\partial^2 V}{\partial z^2}=0$$

The peripheral potentials can be shown to be given by:

$$V_P=\frac{\rho I}{\pi h}\ln\frac{r_1}{r_2}$$

where ρ is the resistivity, I the current, r_1 and r_2 the distances to the point P on the circumference from the points of current injection, and h is the thickness of the disc.

For this homogeneous case we can determine the potential V_i as shown in figure 3.13. Now the conducting material enclosed by two isopotentials may be considered as forming a resistance and in figure 3.13 there are 8 of these resistances which are in series. If resistivity ρ_i is greater than ρ then V_i will be greater than in the homogeneous case. If If ρ_i is less than ρ then V_i will be less than in the homogeneous case. It is therefore intuitively

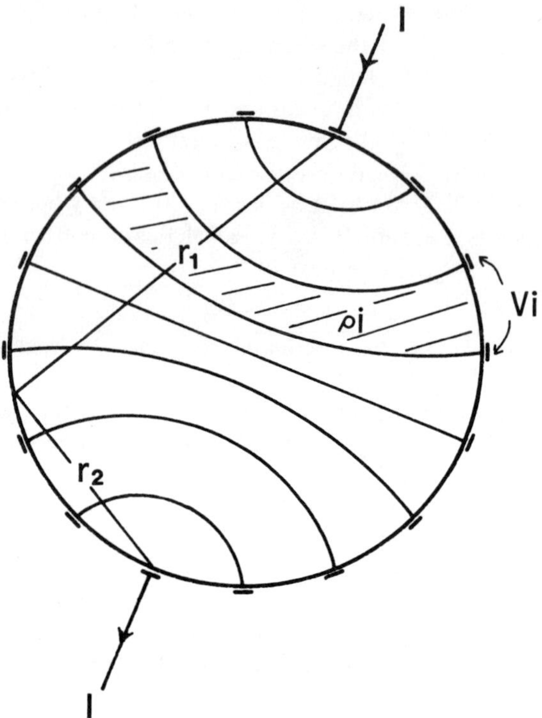

Figure 3.13 The isopotentials to be expected when a current I is applied across a homogeneous disc of material.

reasonable to interpret a high peripheral potential gradient as arising from an increased resistivity in the image elements contained between the two isopotentials which terminate at the points between which the peripheral potential gradient was determined.

We are therefore asserting that a high value of V_i is associated with a high value of ρ_i. If this is the case then it is reasonable to back-project the ratio of the measured peripheral potential gradient to the calculated potential gradient along the isopotential which terminates at this point on the periphery. This back projection may be carried out for every adjacent pair of electrodes on the periphery. For the 16-electrode system shown there are 120 pairs of electrodes through which current can be driven and 1920 peripheral gradient measurements which can be made. n electrodes will yield $n^2(n-1)/2$ peripheral values for back projection. However, not all of these values are independent. By driving current through the n adjacent pairs of electrodes and in each case recording n peripheral gradients then all other potentials which will result from any pair of drive electrodes can be obtained by superimposition. There are thus only n^2 independent measurements. In addition, reciprocity, whereby the peripheral potentials recorded from electrodes p, q when current is driven through r, s will be the same as when recording from r, s and driving p, q, reduces the number of independent measurements to $n^2/2$. If allowance is also made for the fact that a peripheral gradient cannot be measured from the drive electrodes then the number of independent measurements is reduced to $n[(n-3)/2]$.

The system of back projection along curved isopotentials is difficult to handle mathematically. Certainly the techniques covered in section 3.1.2 of volume 1 which considered systems of either parallel or divergent rays are not easily adapted to the new situation. If the isopotential which intersects point (r, θ) in the image for a particular drive electrode configuration D_n is given by

$$I = I(D_n, r, \theta)$$

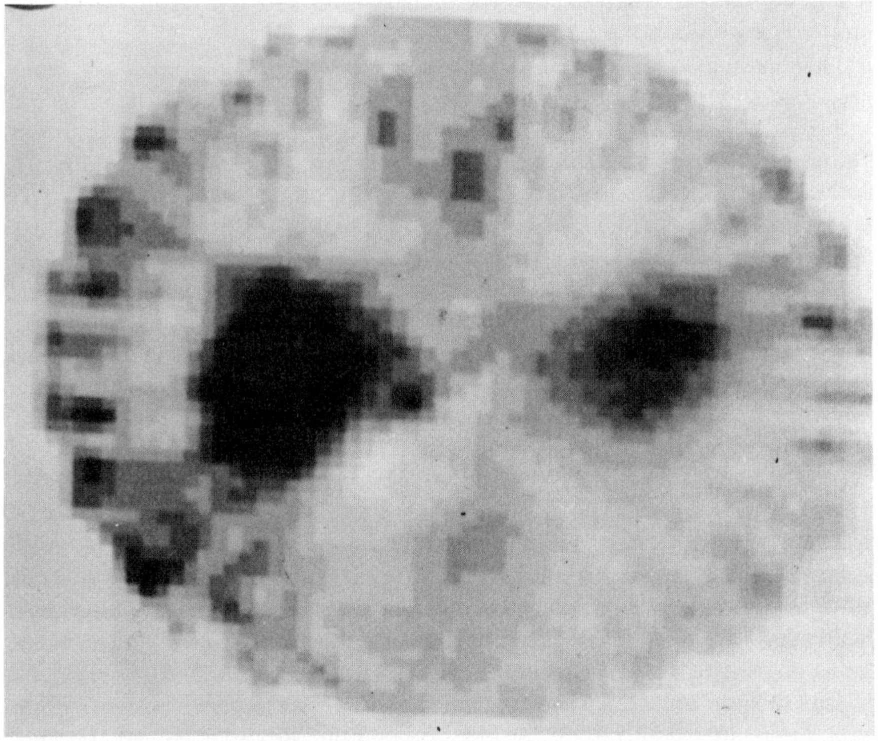

Figure 3.14 A two-dimensional reconstruction of resistivity from slabs of muscle and bone placed in a 150 mm diameter shallow pool of saline. Expected resistivities are: saline, 1.5 Ωm; muscle (on the left) 3 Ωm and bone > 40 Ωm.

and this intersects the periphery at point (R, ϕ) where the expected potential gradient (homogeneous medium) is $U_e(R, \phi)$ and the measured potential gradient is $U_m(R, \phi)$, then we can back-project along I to give the image points

$$f(s, \psi) = \Big|_{\text{along} I} \frac{U_m(R, \phi)}{U_e(R, \phi)}$$

This is carried out for all points in the image and for all values of D_n. The total number of back projections is thus the number of pixels multiplied by the number of drive electrode combinations D_n.

To carry out this system $U_m(R, \phi)$ must be either measured or interpolated for all values of ϕ. Alternatively the back projection can be started from the points where $U_m(R, \phi)$ is measured and then along the associated isopotentials. This method will yield $n^2(n-1)/2$ back projections where n is the number of peripheral electrodes. Intuitively the second method seems less likely to give uniform image reconstruction as each pixel will not be intersected by the same number of isopotentials.

This method of image reconstruction has the major advantage over those described in the previous section that it uses data which can be collected with good accuracy. No electrode guarding, with its associated major practical limitations, is used and the measured values are peripheral potentials, not currents. Peripheral potentials can be recorded to an accuracy of 1% over at least a 40 dB dynamic range and, if the recording system has a sufficiently high input impedance, are not affected by electrode contact impedance. Exit currents are very much dependent on electrode contact impedance which is usually much greater than tissue impedance.

The proposed method of back projection gives an image which contains blurring due to back projection, in addition to that which results from the fact that a point conductivity change will give rise to changes in several peripheral potentials and also to the fact that the isopotentials have only been determined for the homogeneous case. Some of the blurring can be corrected with appropriate filtering of both the peripheral potential profiles and the back projection image (see chapter 3 of volume 1).

Figure 3.14 shows the result of the above method applied to discs of meat (muscle) and bone placed in a shallow 15 cm diameter dish with 16 peripheral electrodes. The single back projection resolves the objects and ranks then in terms of their expected resistivities.

This method of reconstruction lends itself to an iterative procedure, where back projection is followed by recalculation of the isopotentials and expected peripheral potentials for the back projected image. This forward projection involves solving Laplace's equation for the given spatial distribution of resistivity.

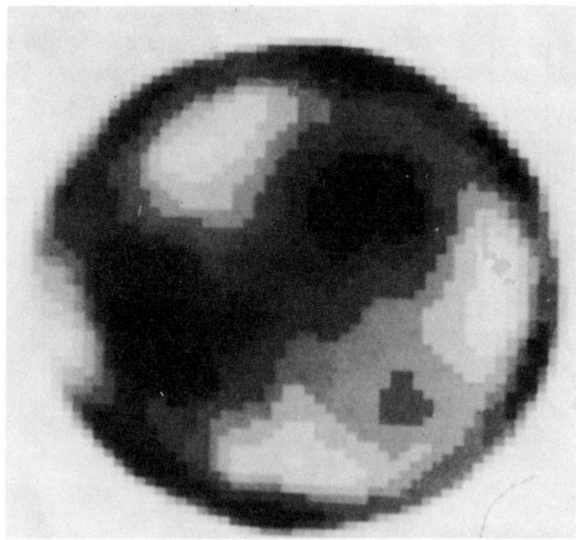

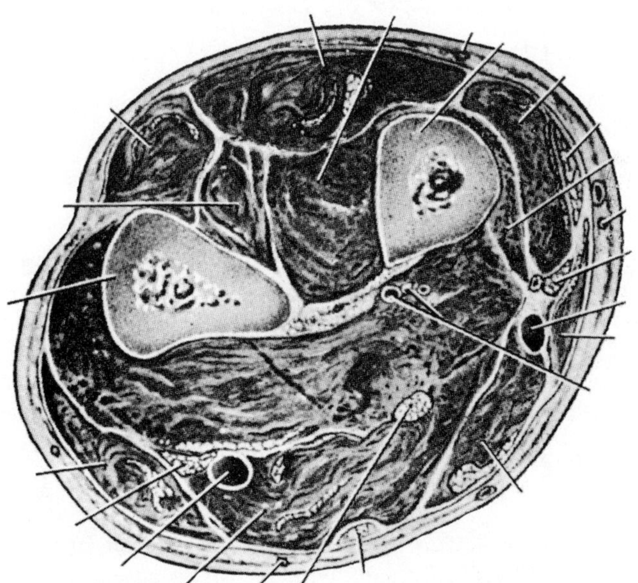

Figure 3.15 A two-dimensional reconstruction of three-dimensional data collected from 16 electrodes on a human forearm is shown at the top, and at the bottom the expected distribution of tissues. It appears that the bones and possibly the major blood vessels are resolved.

In principle the method outlined can also be applied in three dimensions where back projection would be along isopotential surfaces. Figure 3.15 shows the result of a reconstruction from data collected at 50 kHz from a human arm in which some structure can be resolved. Sixteen electrodes were used. However, this image is not the result of 3D reconstruction, which requires formidable computing power, but a 2D reconstruction followed by image filtering.

References

1. R. Pethig, *Dielectric and Electronic Properties of Biological Materials*, John Wiley and Sons, Chichester (1979).
2. D. A. Witsoe and E. Kinnen, *Med. Biol. Eng.* **5** (1967) 239–248.
3. L. A. Geddes and L. E. Baker, *Med. Biol. Eng.*, **5** (1967) 271–293; L. A. Geddes, *IEE Med. Elec. Monographs*, 18–22 (1976) eds. Hill and Watson, 42–72.
4. H. C. Burger and R. van Dongen, *Phys. Med. Biol.* **5**(4) (1961) 431–447.
5. L. Cromwell, F. J. Weibell, and A. E. Pfeiffer, *Biomedical Instrumentation and Measurements* (2nd edn.) Prentice Hall, New Jersey (1980).
6. R. Plonsey, *Bioelectric Phenomena.* McGraw-Hill, New York (1969).
7. D. Witsoe, University of Minnesota MS thesis (1966).
8. R. Rush, *J. Res. National Bureau of Standards Section C*: Vol. 66C, No. 3 (1962) 217; S. Rush, J. A. Abildskov and R. McFee, *Circ. Res.* **12** (1963) 40–50.
9. S. Stefanesco, C. Schlumberger, and M. Schlumberger, *J. Phys.* **1** (1930) 133.
10. D. W. Hill and S. N. Mohapatra, *IEE Medical Electronics Monographs*, Peter Peregrinus/IEE, London (1977) 23–27.
11. B. H. Brown, W. I. J. Pryce, C. D. Baumber and R. G. Clark, *Med. Biol. Eng.* **13** (1975) 674–682.
12. D. W. Hill and A. M. Dolan, *Intensive Care Instrumentation*, Academic Press, New York (1976).
13. R. Plonsey and R. Colin, *Med. and Biol. Eng. and Comput.*, **15** (1977) 519–527.
14. R. P. Henderson and J. G. Webster, *IEEE Trans. Biomed. Eng.*, Vol. BME-25 (1978) 250–253.
15. H. Schomberg and M. Tasto, *Reconstruction of Spatial Resistivity Distribution from Resistance Projections*, Philips GmbH, Hamburg, MS-H, 2715/81, (1981).
16. L. R. Pryce, *IEEE Trans. Nucl. Sci.* Vol. NS-26, No. 2 (1979) 2736–2739.
17. K. A. Dines and R. J. Lytle, *Geophysics* **46**, No. 7 (1981) 1025–1036.
18. H. J. Schneider, *On the Ambiguity of Impedance Tomography*, (in German), Philips GmbH Forschungslaboratorium, Hamburg, LB 448/79 (1979).
19. R. H. T. Bates G. C. McKinnon and A. D. Seagar, *IEEE Trans. Biomed. Eng.*, Vol. BME-27 (1980) No. 7, 418–420.

Section 3.4.3, and in particular the method used to produce figures 3.14 and 3.15, is the subject of the following provisional patent application: B. H. Brown, D. C. Barber and I. L. Freeston, Patent Application No 8212676, Title-Tomography, April 1982.

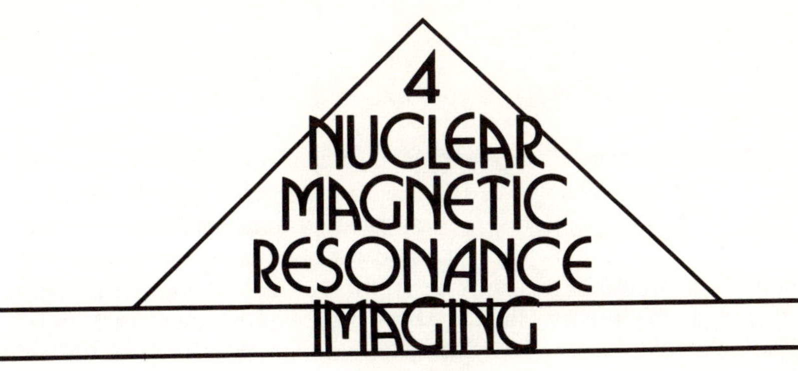

P.S. Allen

4.1 Introduction

As an experimental technique in the physical sciences, nuclear magnetic resonance (NMR) has a long and successful history reaching back to the 1940s. Since that time the developments in electronics and particularly in computing, together with a growing understanding of the phenomenon itself, have permitted a continual increase in the size and complexity of the problems amenable to NMR methods. In recent years, the application of NMR has expanded its domain from the basic physics and chemistry into the more complex life sciences and imaging is just one of the applications in this area. The basic principles of NMR itself are covered in numerous textbooks, pre-eminent amongst which is probably the book by Abragam (1). Accounts of NMR imaging are not so common. A comprehensive book has been written by Mansfield and Morris (2), and at the other extreme, an outline for the lay scientist has been covered by a *Scientific American* article by Pykett (3).

The principal objective of this contribution is to describe in detail the physical principles underlying some of the different methods currently used in NMR imaging. The details of the medical and biological applications of these applications are not included here. However, before introducing in section 4.3 the techniques which have enabled NMR to become an imaging modality, an account is given in section 4.2 of those basic concepts of NMR which are critically involved in the imaging application. For example, an understanding of the mechanism by which the nuclear signal is generated, of the idea of the rotating co-ordinate frame and of the manner in which the nuclear signal is modified by various relaxation processes, are all essential prerequisites to an appreciation of NMR imaging. In the final section, section 4.4, the question of discrimination in NMR images is addressed. In contrast with imaging using X-rays, the NMR image intensity is not solely a function of the nuclear density at each site in space. It can be made to reflect characteristics, namely relaxation rates, which depend heavily on the local molecular environment and as a result NMR imaging possesses a built-in discrimination mechanism.

When NMR imaging is employed in medicine or biology, the nucleus which is almost always exploited is the proton. This reflects the large proportion of hydrogen atoms in biological systems and also the high suitability of the proton for NMR work. Because of this, the following sections focus entirely on proton imaging. A description of the use of other nuclei in NMR imaging, which is very limited at this time, can be found in reference (2). In biological systems the hydrogen atoms find themselves most often in water molecules. Moreover, the molecular mobility of the water molecules makes the resonance characteristics of their proton nuclei very suitable for imaging purposes. Many of the hydrogen atoms in large molecules like proteins are not so well endowed. For example if protons exist in a fairly rigid molecular environment, then their relaxation rates are likely to be such as to inhibit their NMR signal generation. As a result, an NMR image emphasizes the more mobile protons in the system and in particular those in water molecules. NMR is therefore very efficient at illustrating soft tissues and the differences between them. In this respect, it is complementary to X-ray methods which detect the rigid structures of the body most easily.

As NMR imaging has developed, many methods have been proposed for obtaining the raw NMR data. It is convenient to divide these methods into categories which reflect the region from which that raw data is obtained. For example, the sensitive point category and the line scan category, which were devised early in the development of NMR imaging, collect their data point by point or line by line respectively. Because of this they tend to be rather slow and fail to take full advantage of all the NMR signal available from a two-dimensional slice. The current trend is to exploit methods in the planar imaging category, which when generating an image, use the NMR signal from the whole of a two-dimensional slice. It is this latter category of methods which we shall discuss in section 4.3. A future trend towards a volume imaging category has begun and though it is mentioned briefly in section 4.3.3 it is not discussed in detail in this contribution.

4.2 Some basic concepts involved in NMR imaging

4.2.1 *Nuclear magnetic resonance*

Not all nuclear species are admitting of nuclear magnetic resonance. This is because not all nuclear species exhibit a nuclear magnetic moment. To exhibit a magnetic moment, a nucleus must possess a spin angular momentum $\mathbf{I}\hbar$ (where $\mathbf{I}$ is the nuclear spin quantum number) which in turn arises from the vector addition of the nucleon angular momenta only in those nuclei which are composed of odd numbers of protons and/or neutrons. Associated with this spin angular momentum there exists a concomitant magnetic moment of component μ, parallel to the angular momentum vector and whose magnitude is given by

$$\mu = \gamma \cdot \mathbf{I}\hbar \tag{4.1}$$

where γ is called the nuclear gyromagnetic ratio. γ depends on the nuclear structure and is unique for each nuclear species. For nuclear species in general, the finite values which **I** can take include all integral and half integral values from $\mathbf{I} = \frac{1}{2}$ to $\mathbf{I} = 8$. However, we are fortunate in that so far as NMR imaging is concerned, the principal nucleus of interest is the proton, for which $\mathbf{I} = \frac{1}{2}$, and for which therefore, the nuclear resonance description is the simplest. As a result, we shall confine the remainder of this discussion to the case of $\mathbf{I} = \frac{1}{2}$.

When a collection of nuclei of $\mathbf{I} = \frac{1}{2}$ is situated in a uniform static magnetic field $\mathbf{B}_0$, the nuclei distribute themselves amongst the two stationary energy levels corresponding to parallel and antiparallel alignment of the magnetic moments $\boldsymbol{\mu}$ with the magnetic field $\mathbf{B}_0$. This population distribution conforms to Maxwell-Boltzmann statistics and figure 4.1 illustrates the populations and the energies of the nuclei in these two energy states. If such a system is now irradiated with electromagnetic radiation whose angular frequency, ω_0, satisfies the resonance condition

$$\hbar\omega_0 = (E_2 - E_1) = 2\mu B_0$$

giving, together with equation (4.1),

$$\omega_0 = \gamma B_0 \tag{4.2}$$

then the magnetic vector of the electromagnetic field can interact with the nuclear moments and stimulate transitions of the nuclei between the stationary energy states. Equation (4.2) is the basic equation of NMR, giving the resonant radiation frequency in terms of the applied static magnetic field

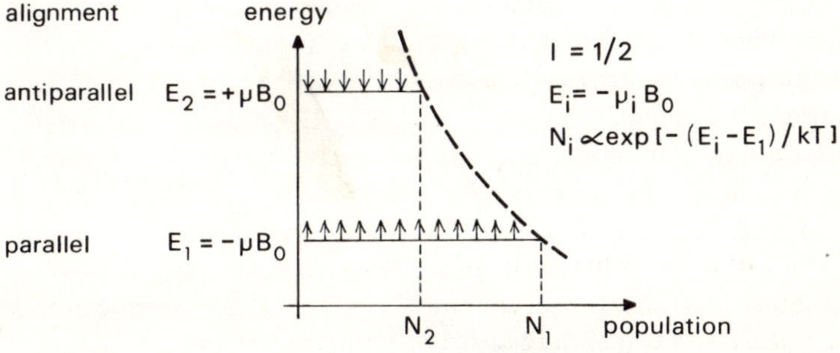

Figure 4.1 An illustration of the distribution of parallel and antiparallel nuclear magnetic moments for a collection of spins $I = \frac{1}{2}$ in a static magnetic field B_0.

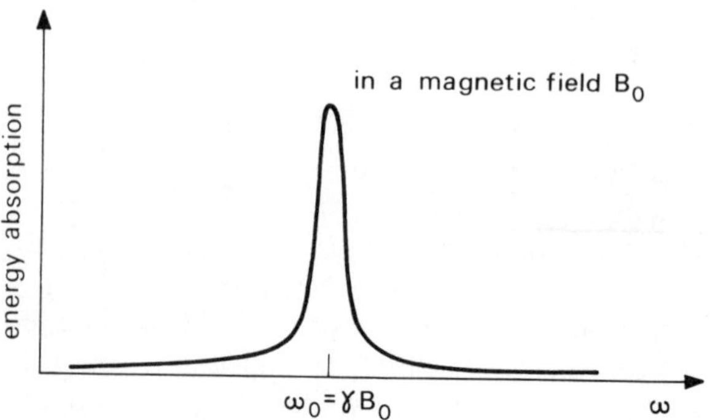

Figure 4.2 A diagrammatic representation of the resonant absorption of energy from an electromagnetic field by a collection of nuclear spins in a static magnetic field B_0. The intensity of the energy absorption is proportional to the number of nuclei in the collection.

B_0 and the nuclear species through γ. The resonant frequency is typically in the radio-frequency (r.f.) region of the electromagnetic spectrum when $\mathbf{B}_0$ is a fraction of one tesla, e.g. for protons when $B_0 \sim \frac{1}{4}T, \omega_0/2\pi \sim 10\,\text{MHz}$. To observe this resonance phenomenon, as illustrated in figure 4.2, one can weakly irradiate the nuclear system at a constant frequency ω_0 and measure its absorption of energy from the radiation field, while slowly sweeping the static magnetic field. When the condition of equation (4.2) is satisfied, the absorption of energy undergoes a sharp increase as a significant fraction of the nuclei with energy E_1 take energy from the radiation field, in order to flip their orientation and become antiparallel to B_0.

The simple description of the magnetic resonance phenomenon as presented above, in terms of energy levels and transitions caused by a weak r.f. perturbation, is of only limited application in the wider context of NMR as a whole. A more useful approach is to examine the behaviour of the net magnetization, $\mathbf{M} = \sum \boldsymbol{\mu}_i$, which is produced by the nuclear magnetic moments when they are aligned in each of their respective energy levels in the static magnetic field $\mathbf{B}_0$. This approach enables us not only to account for the nuclear response to strong r.f. fields, but also to provide a model for nuclear relaxation phenomena.

The alternative magnetization approach is illustrated in figure 4.3, where the nuclei in each of the two energy states can be considered to give rise to a resultant magnetic moment per unit volume (or magnetization) either parallel or antiparallel to $\mathbf{B}_0$, which in turn defines the z axis of the laboratory co-ordinate system. Since in thermal equilibrium there is an excess of nuclei in the lower energy state, the parallel magnetization will be the larger of the two and the thermal equilibrium situation will correspond to a net resultant magnetization parallel to $\mathbf{B}_0$ along the positive z axis. More generally, the behaviour of the

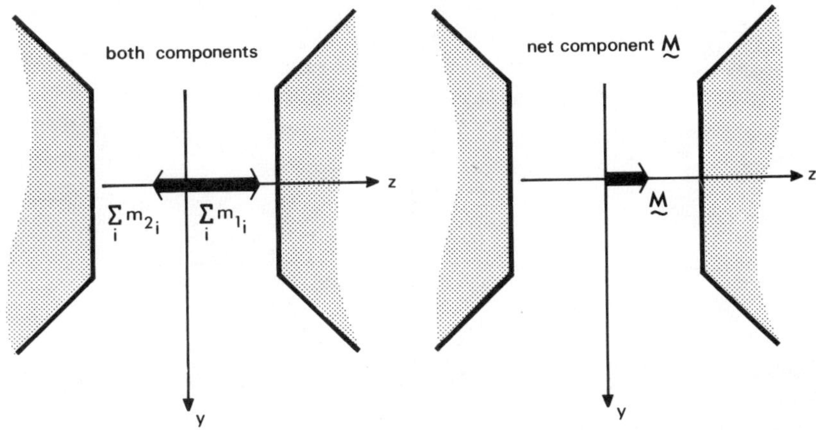

Figure 4.3 A schematic diagram showing the magnetization vectors $\sum_i m_{1_i}$ and $\sum_i m_{2_i}$ resulting from the nuclear moments in each of their energy states E_1 and E_2 respectively, together with the resultant net magnetization **M** representing the excess population in the E_1 state.

nuclear spin system may be analysed in terms of this resultant magnetization vector, **M**, which, although parallel to $\mathbf{B}_0$ in thermal equilibrium, undergoes orientation and magnitude changes as a result of the influence of various perturbations. For example, suppose we consider first the effect on the thermal equilibrium magnetization of a strong burst of r.f. radiation. If this radiation satisfies the resonance condition of equation (4.2) and is in addition circularly polarized so that its magnetic vector $\mathbf{B}_1$ is always perpendicular to $\mathbf{B}_0$, then **M** undergoes a precessional motion in which it effectively rotates around $\mathbf{B}_0$ at the angular frequency $\omega_0 = \gamma B_0$ and at the same time around $\mathbf{B}_1$ at an angular frequency $\omega_1 = \gamma B_1$. To establish the general existence of precessional motion around a magnetic field, **B**, one needs to recall from classical electromagnetism that a torque $\mathbf{M} \wedge \mathbf{B}$ is exerted on a magnetization, **M**, when it is placed in that field. This torque is equal to the rate of change of angular momentum $d(\mathbf{I}\hbar)/dt$ and hence the motion of **M**, which is itself related to $\mathbf{I}\hbar$ by equation (4.1), is described by

$$\dot{\mathbf{M}} = \gamma\, \mathbf{M} \wedge \mathbf{B} \tag{4.3}$$

Verification of the nature of the precessional motion occurring under the influence of an r.f. pulse can be obtained by setting up the equation of motion of **M**, under the influence of two fields, $\mathbf{B}_0$ and $\mathbf{B}_1(t)$, and then by transforming it into a coordinate frame rotating about z at the same frequency as the r.f. radiation. This piece of analysis can be found in several textbooks on NMR (1) and attractive diagrams of the precessional motion can be found in a *Scientific American* article by Pykett (3). Figure 4.4 illustrates the resulting

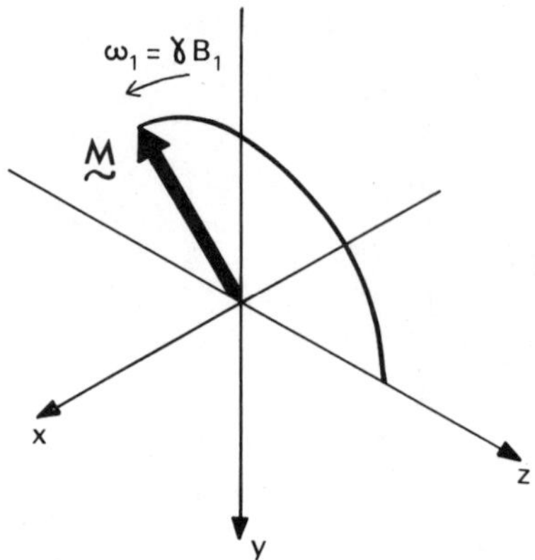

Figure 4.4 A diagram to illustrate the rotation in the rotating frame of the net magnetization **M** around the polarization direction of the r.f. magnetic field, during the irradiation.

behaviour in the rotating coordinate frame. If the burst of r.f. radiation is terminated exactly at a time t such that

$$\omega_1 \cdot t = \pi/2$$

then, following the burst, **M** will be left in the xy plane. If on the other hand, the burst is allowed to continue for a time t', such that

$$\omega_1 \cdot t' = \pi$$

then **M** will be deposited along the negative z axis at the end of the r.f. pulse. Bursts of r.f. power of these two specific lengths are important in NMR and they are called respectively, 90° and 180° pulses. They are illustrated in figure 4.5, where the relationship with the population distribution is also shown.

Following the termination of an r.f. pulse, the only magnetic field acting on the nuclear spin system is $\mathbf{B}_0$ and therefore any magnetization component not parallel to $\mathbf{B}_0$ will precess around the z axis at the angular frequency ω_0. This precessional motion of the magnetization, in the absence of r.f. stimulation, provides a means for detecting the nuclear signal. For example, if a search coil were placed around the nuclear system, with its axis in the xy plane, the rotating xy component of the magnetization would induce in the search coil an EMF, oscillating at ω_0 and whose magnitude was proportional to that xy magnetization. This oscillating EMF is the nuclear signal. It is often called the Free Induction Decay or FID and in the absence of any relaxation effects (and these are to be discussed in section 4.2.2) it would continue indefinitely at a constant amplitude.

At this point it is useful to bring together the two approaches to NMR which have been presented above. One, in terms of transitions between energy

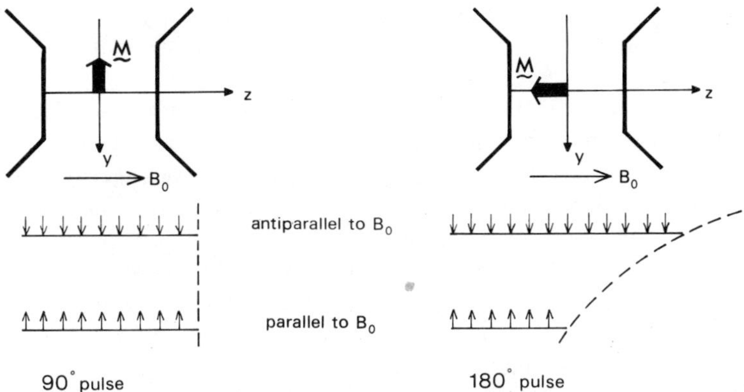

Figure 4.5 An illustration of the effects of a 90° pulse on the one hand and a 180° pulse on the other. These effects are shown both from the point of view of the populations and from the point of view of the net magnetization.

levels, gave rise to a nuclear signal which was a delta function in frequency space and located according to the resonance condition of equation (4.2). The other in terms of resultant magnetizations precessing under the influence of magnetic fields, gave rise to a nuclear signal which was a continuous sinusoid in the time domain. These two methods of observing the nuclear signal are equivalent and the relationship between them is the Fourier transform (4). By application of this transform, a frequency spectrum can always be derived from the time domain FID waveform and vice versa. When performing imaging experiments the data is invariably obtained as a time domain signal resulting from pulsed r.f. stimulation. However, in order to generate an image the frequency spectrum is usually required and so the step of taking the Fourier transform to go from one to the other is an essential one in NMR imaging.

4.2.2 *The nuclear relaxation times T_1 and T_2*

In practice any magnetization component which exists in the *xy* plane following an r.f. pulse will not precess around the *z* axis indefinitely. Because of interactions which take place on an atomic scale, the magnetization vector will undergo two distinct relaxation processes simultaneously. One concerns the recovery, towards equilibrium, of the Boltzmann populations of the energy states and hence the recovery of the magnetization along the *z* axis and parallel to $\mathbf{B}_0$. This process is called spin-lattice relaxation and is characterized by a time constant T_1. The second relaxation process is called spin-spin relaxation and it is characterized by a time constant T_2. This process concerns the behaviour, within the xy plane, of the components of the magnetization vector which find themselves in that plane at the termination of the r.f. pulse.

Spin-lattice relaxation (T_1). When, as a result of r.f. irradiation, the nuclear spin system is disturbed from its thermal equilibrium, there is a corresponding change in the relative populations of the two nuclear spin energy states. To re-establish the thermal equilibrium populations in the absence of any r.f. irradiation, transitions are required to take place, which in turn require a mechanism providing both in interactive coupling to the nuclear spins and an energy reservoir to allow energy exchange to and from the nuclear system. In molecular systems the most powerful spin-lattice relaxation mechanism is due to the dipole-dipole interaction and this interaction will be discussed in more detail in the Appendix, section A1. The interactive coupling in the dipole-dipole interaction is the magnetic coupling between two neighbouring magnetic dipoles, usually nuclear dipoles, and the energy reservoir is the thermal bath of motional kinetic energy within the molecular system. For instance a nucleus, whose neighbour undergoes random thermal motion, will experience a fluctuating magnetic field from the nuclear magnetic dipole of that neighbour. This fluctuating field will, no doubt, have a Fourier component at a resonance frequency of the original nucleus, and this component will be capable of inducing a transition of that nucleus between its stationary energy states. When a transition occurs nuclear spin energy is converted to motional kinetic energy or vice versa. In this manner, the nuclear spin system and the thermal reservoir of molecular motion can equilibrate their energies and since the thermal capacity of the thermal reservoir is overwhelmingly greater than that of the spin system, the net result is always to bring the spin system into thermal equilibrium with the thermal bath or lattice, as it is usually called in magnetic resonance terminology.

To evaluate the form of the recovery toward equilibrium, it is easiest to consider the populations of the two nuclear spin states, denoted by N_1 and N_2 and illustrated in figure 4.1. A measure of the deviation of the nuclear spin system from thermal equilibrium is the population difference n given by

$$n = N_1 - N_2$$

Now if the system has been disturbed from equilibrium and is being allowed to recover we can postulate relaxation transitions of probability per second $W_\uparrow$ and $W_\downarrow$ for the upward and downward transitions respectively. During the recovery, the rate of change of the ground state population is

$$\frac{dN_1}{dt} = N_2 W_\downarrow - N_1 W_\uparrow \tag{4.4}$$

but to analyse the return to equilibrium we require a differential equation for n rather than for N_1. Changing the variables in equation (4.4) from N_1 and N_2 to n and the total number of spins $N = N_1 + N_2$, we have

$$\frac{dn}{dt} = N(W_\downarrow - W_\uparrow) - n(W_\downarrow + W_\uparrow)$$

which can be written in the form

$$\frac{dn}{dt} = (A - n)k \tag{4.5}$$

where $A = N(W_\downarrow - W_\uparrow)/(W_\downarrow + W_\uparrow)$ and $k = (W_\downarrow + W_\uparrow)$. Before discussing the solution to equation (4.5), it is useful to identify the nature of the constant A. The relaxational transition probabilities $W_\uparrow$ and $W_\downarrow$ are not equal, as are the probabilities of transitions stimulated by the r.f. irradiation. The reason for this is that the thermal reservoir of the lattice is involved in the relaxation transitions and the rate of transitions depends not only on the matrix elements of the dipolar coupling but also on the probability that the reservoir will be in a state that permits an exchange of energy. We can obtain the relationship between $W_\downarrow$ and $W_\uparrow$ by remembering that in thermal equilibrium a steady state exists, so that the equilibrium populations $N_1(\text{eq})$ and $N_2(\text{eq})$ are connected by

$$N_1(\text{eq}) \cdot W_\uparrow = N_2(\text{eq}) \cdot W_\downarrow$$

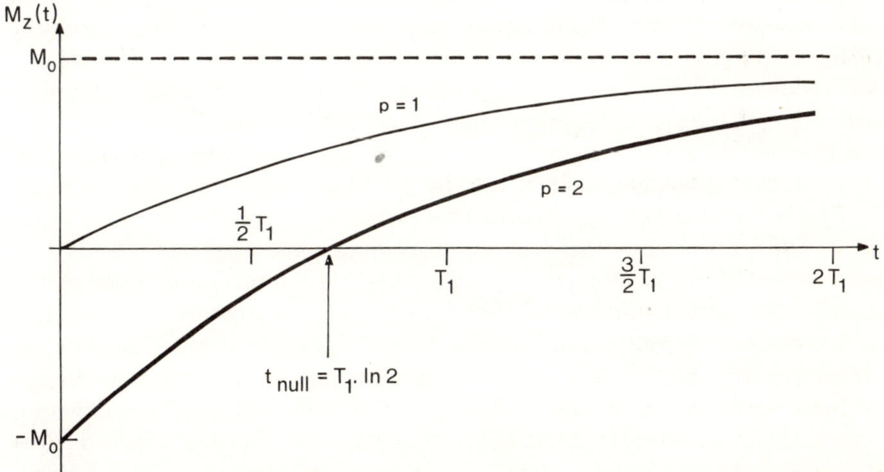

Figure 4.6 A graph representing the recovery of the z magnetization $M_z(t)$ following either a 90° r.f. pulse or a 180° r.f. pulse. The time axis is scaled in units of T_1 to emphasize the time taken to regain thermal equilibrium.

so that

$$\frac{W_\uparrow}{W_\downarrow} = \frac{N_2(\text{eq})}{N_1(\text{eq})} = \exp\{-(E_2 - E_1)/kT\} \tag{4.6}$$

By means of equation (4.6) one can show that $A = n(\text{eq})$, the equilibrium population difference. Solving equation (4.5) and substituting the boundary conditions corresponding to stimulating 90° or 180° pulses one obtains

$$n = n(\text{eq})\{1 - p\exp(-kt)\} \tag{4.7}$$

where $p = 1$ following a 90° pulse stimulation, and 2 following a 180° pulse. Thus the population difference displays an exponential recovery and the time constant T_1 is defined by

$$\frac{1}{T_1} = k = (W_\uparrow + W_\downarrow) \tag{4.8}$$

Remembering that the net magnetization in the z direction, M_z, is proportional to n, we can see that the recovery of M_z, following a 90° or a 180° pulse, will be as illustrated in figure 4.6 if it is driven by a single spin-lattice relaxation mechanism.

Spin-spin relaxation (T_2). Following the termination of a burst of r.f. irradiation, the general magnetization vector **M** can be resolved into a z component and an xy component. The concept of spin-spin relaxation provides a means to describe the behaviour of the xy component alone. In the ideal case where all the nuclei experience exactly the same uniform static magnetic field $\mathbf{B}_0$ and do not interact with each other or their environment, the magnetization vector in the xy plane will precess indefinitely about $\mathbf{B}_0$ at the angular frequency ω_0 given by equation (4.2). In practice nuclear spins do interact both with neighbouring nuclear spins and with their electronic environment. The consequence of these interactions is in general twofold.

The first is to cause the magnetic fields which the nuclei experience to vary from site to site. In other words, a local magnetic field distribution is superimposed on the uniform applied field $\mathbf{B}_0$. As a result the nuclear energy levels suffer a broadening, which in turn, leads to a corresponding broadening of the resonance spectrum in frequency space. From the magnetization view, the important point is that individual nuclear moments will not all precess at the same frequency. Those with local fields which reinforce $\mathbf{B}_0$, will precess faster than ω_0, whereas local fields which oppose $\mathbf{B}_0$ will cause nuclear moments which experience them to process more slowly than ω_0. The net effect of this will be a distribution in precessional rates and hence the xy magnetization vector will fan out as the component nuclear moments get out of phase with each other. The fanning out of the magnetization vector will result in a reduction of its effective magnitude and therefore a decay of the EMF it induces in the receiver coil, hence the name free induction decay or FID which is given to this form of nuclear signal.

The second effect of the interactions of the nuclei with their surroundings is to limit the lifetime of the nuclear spin states, either by spin-lattice-relaxation processes or by the flip-flop mechanism in the dipole-dipole interaction (see Appendix, A1). In either event, the limitation on the lifetime interrupts the rotational coherence of the nuclear magnetic moments and also leads to a lifetime broadening of the nuclear energy levels. It thus gives rise to the same sort of effect as the distribution in local fields, in both the frequency domain (i.e. a broadening of the spectrum) and the time domain (i.e. a decay of the induced EMF).

Although the two consequences of environmental interactions have similar effects, detailed differences in behaviour do arise depending on which of the two is the dominant one. For example, if the nuclear environment is static, as in solids, the local magnetic field distribution is large, the frequency spectrum will be very broad and often Gaussian in character and the FID will be particularly short and non-exponential in time. The frequency spectrum and the FID will, however, still be related by the Fourier transform operation. In contrast, if the nuclear environment is very mobile, as is found in liquids, the local field distribution tends to average to zero because of the motion of the magnetic dipoles. In this extreme narrowing limit only the weaker lifetime broadening remains, is usually Lorentizian in character and the corresponding decay of the nuclear precession signal is therefore long and exponential. In such circumstances, it is meaningful to define a characteristic time T_2 for the decay of the free precession signal. In this limit T_2 can approach the value of T_1 but never exceed it.

4.2.3 *The phenomenological equation of Bloch*

When the interactions between neighbouring nuclei are strong, as is usually the case in molecular solids, the timescale of the FID is very short (20–200 μs) and the practical requirements of NMR imaging become particularly severe. Because of this, the technique finds its greatest utility in situations where the effective interactions between neighbouring nuclei are weak and the FIDs are correspondingly long (10–100 ms) and often exponential in character. Such situations occur in the liquid-like systems found in medicine and biology. In the weakly interacting situation, the behaviour of the nuclear magnetization can be described by a pseudo-classical model due to Bloch (5), who made the assumption that the behaviour of $\mathbf{M}$ resulting from relaxation could be superimposed on the motion of the free spin magnetization in a magnetic field, as given by equation (4.3). Thus by combining the torque on

M due to the applied magnetic fields, with an exponential spin-spin relaxation of its xy components M_x and M_y and the spin-lattice recovery of the xz component as described by equation (4.5), we obtain the Bloch equation in the laboratory frame of reference, i.e.

$$\frac{d\mathbf{M}}{dt} = \gamma \mathbf{M} \wedge \mathbf{B} - \frac{\{M_x \cdot \mathbf{i} + M_y \cdot \mathbf{j}\}}{T_2} + \frac{\{M_z - M_0\}\mathbf{k}}{T_1} \tag{4.9}$$

where M_0 is the equilibrium magnitude of M_z, and **i**, **j** and **k** are unit vectors parallel to the laboratory frame cartesian axes. Solutions of the Bloch equation in its various forms enable one to account for the behaviour of the magnetization either during r.f. irradiation or in between pulses. For example, for imaging experiments in general **B** will be composed of a static homogeneous field $\mathbf{B}_0$, an orthogonal r.f. field $\mathbf{B}_1(t)$ and a linear magnetic field gradient $\mathbf{B}_g(x, y, z)$. The significance of the field gradient will be introduced lated in section 4.3.1. During intense r.f. pulses $\mathbf{B}_g(x, y, z)$ is usually negligible and can be omitted, whereas between the pulses it is $\mathbf{B}_1(t)$ that is zero. We shall therefore have at least two forms of the Bloch equation, one with a time-dependent **B** and one with a time-independent **B**.

The method of solution (6) of the Bloch equation is a matter of preference. One way is to use matrix techniques (7, 8) which are useful in more general situations but which can become more involved than is necessary in very simple cases. The appendix (A2) illustrates the matrix method by performing the transformation to a reference frame rotating with the r.f. field, in order to obtain a time-independent differential matrix equation which can be solved to give the magnetization behaviour during an r.f. pulse.

4.3 Imaging techniques

4.3.1 *Magnetic field gradients*

The fundamental objective in NMR imaging is to identify individually the nuclear signal coming from each location in space. To fulfil this objective we require a method of encoding spatial information into the nuclear signals and the way this is done is to use well defined magnetic field gradients.

The term 'magnetic field gradient' describes a spatial variation of the magnitude of a magnetic field but not a variation in its direction. The most widely used gradients in NMR imaging are linear gradients and one of these is illustrated in figure 4.7. In the illustration, the gradient is centred at the origin of the co-ordinate system and the length of the arrows represents the magnitude of the gradient field to be added to or subtracted from the main homogeneous field $\mathbf{B}_0$, applied in the z direction. These field gradients endow different locations in space with different magnitudes of applied magnetic

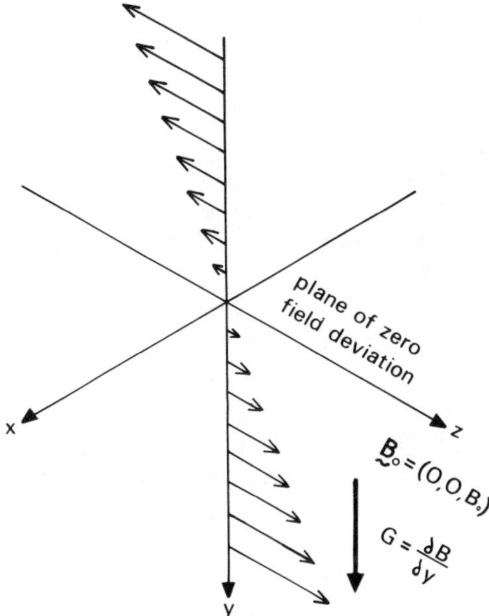

Figure 4.7 A diagrammatic representation of a linear magnetic field gradient in the y direction. The field direction is in the z direction.

field and thereby ensure that the resonance condition of equation (4.2) varies in a controlled manner from place to place, in the sample.

Although it is possible to perform NMR imaging directly in three dimensions (see section 4.3.3), it is more common to define first a slice through the sample and then encode for spatial identification within that two-dimensional slice. When a two-dimensional image is to be obtained, the gradients not only enable the spatial encoding to be achieved within the slice, but they also play a crucial role in the initial definition of the slice itself (see section 4.3.2).

As a first approximation, the superposition of a linear gradient $\mathbf{G}_r$ on to a uniform static field, $\mathbf{B}_0$, has the effect of dividing the sample into elemental slices normal to the gradient direction, $\mathbf{r}$, and each slice, at distance $\mathbf{r}$ from the origin, will have its own resonance condition given by

$$\omega_r = \gamma(B_0 + \mathbf{G}_r \cdot \mathbf{r}). \tag{4.10}$$

As a result of this and following the termination of a burst of r.f. irradiation, any magnetization component normal to the z direction will precess around z at a frequency ω_r and therefore induce an EMF in the receiver coil at this same frequency. This is illustrated for two elemental samples in a y gradient in figure 4.8. By picking out from the overall EMF components at discrete intervals of frequency and then by analysing their respective amplitudes, it is possible to obtain a measure of the number of resonating nuclei at each value of **r**. The process of picking out the individual frequency components and evaluating their amplitudes is carried out by performing the Fourier transform operation on the time-domain EMF waveform. The resulting frequency-domain envelope is then a projection of the nuclear density on to the gradient direction and when generating an image many such projections may be needed (see section 4.3.3).

The generation of linear magnetic field gradients is brought about with current-carrying coil systems. In terms of the coordinate frame in which the z axis is defined by the applied field $\mathbf{B}_0$, the z gradient can be obtained from a pair of coaxial coils whose axes coincide with the field direction. This pair, often called a Maxwell pair, is characterized by counter-rotating currents and a coil spacing which is $\sqrt{3}$ times their radius. The x and y gradients however, cannot be produced quite so easily. They are generated by a system which approximates to a quadrupole arrangement of infinite line currents (9, and the references therein) as shown in figure 4.9. These line currents are orthogonal to both the required gradient direction, r, and the static magnetic field, $\mathbf{B}_0$. Each wire carries a current of identical magnitude and flowing in the same direction. Figure 4.9 illustrates the vector addition of the field

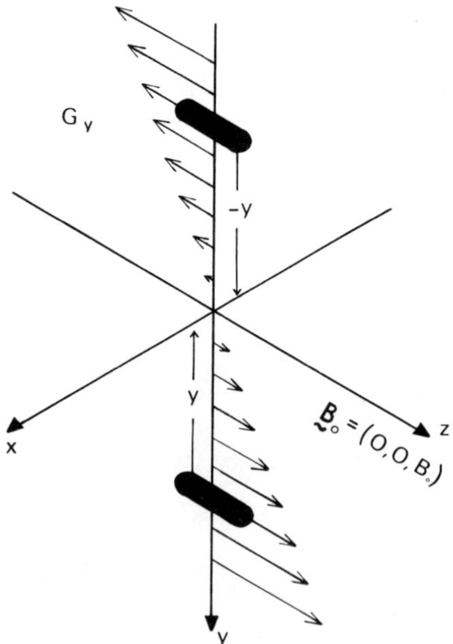

Figure 4.8 (for legend see next page).

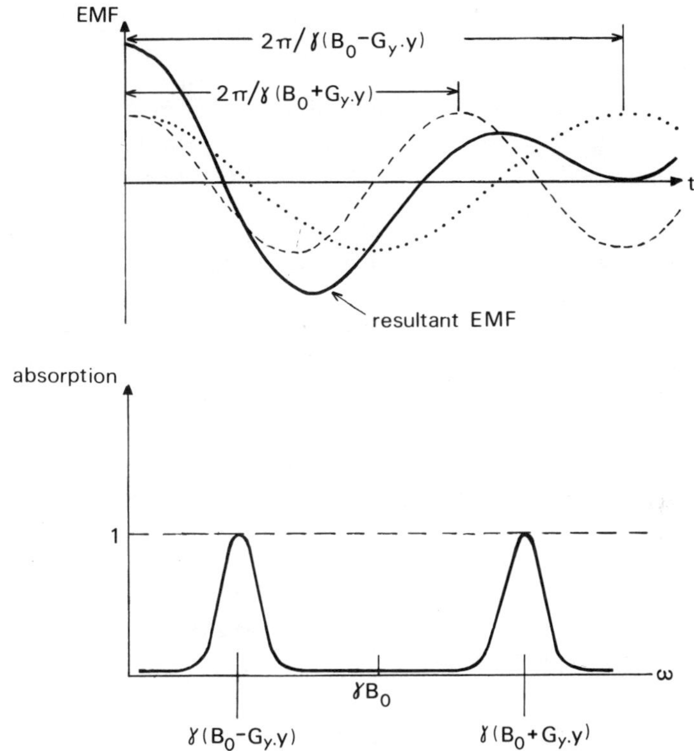

Figure 4.8 An illustrative example of the time and frequency domain signals from two elemental sources of magnetization placed at $\pm y$ in a magnetic field gradient G_y. The resultant EMF arises from the two component cosine signals and when it is Fourier transformed, it produces peaks in the frequency domain at the respective frequencies of the cosine signals. The frequency domain envelope then represents a projection of the magnetization on the y axis.

components from the individual line currents, showing that in the central region the r component of the field cancels but that the z component parallel to $\mathbf{B}_0$ is finite, passes through zero in the $r=0$ plane and reverses direction as r changes sign. To maximize the region of linearity, the ratio z_0/r_0 is adjusted to eliminate the third order gradient. In practice, the line currents cannot be infinite in length and the current-return paths must be carefully designed to minimize their effect on the field direction and the gradient linearity. The detailed configuration of these return paths will be governed by whether the overall geometry is to be parallel planes or cylindrical in character, which in turn is usually governed by the geometry of the main magnet, which may have pole faces or may be cylindrical in form. Because NMR imaging often requires the rapid switching of the field gradients, it is

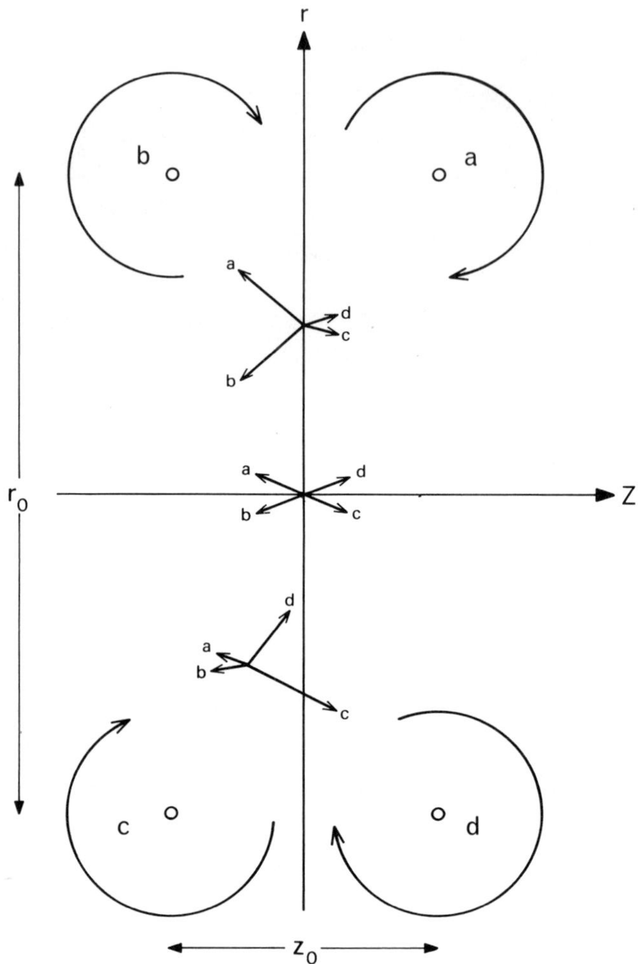

Figure 4.9 A cross-section of the line currents *a*, *b*, *c* and *d* which generate a gradient in the *r* direction. The circles represent the circulatory nature of the magnetic fields due to each wire independently. The three vector diagrams represent the fields due to each of the wires at three different points in the central region.

often important to minimize the inductance of the whole coil assembly and this requirement places additional constraints on the return path geometry (10). To produce the remaining gradient orthogonal to *z*, it is only necessary that the quadrupole coil system be rotated about the *z* axis through an angle of 90°.

4.3.2 *Slice definition*

It was mentioned in section 4.1 that the most common current strategy in NMR imaging is to generate a two-dimensional image from data obtained simultaneously from all locations within a two-dimensional slice through the object. This strategy requires that all the nuclei within such a slice be selectively excited without perturbing the nuclei on either side of the slice. In very simple terms, one can do this by encoding a unique resonance frequency into the slice by means of a linear magnetic field gradient. Then by irradiating the system selectively at that specific resonant frequency only, one can selectively excite that slice. To irradiate selectively over a narrow band of frequencies is not a trivial exercise and the following explanation describes but one way to achieve that end. Although a guide to the shape of the irradiated region can be obtained by considering the Fourier transform or frequency domain spectrum of the irradiation waveform, a better understanding of the process is obtained by examining how the individual magnetization components within the sample are affected by the irradiation sequence.

The behaviour of the magnetization components during irradiation is best visualized in a coordinate frame rotating with the r.f. magnetic field. This is called the *rotating frame* and in the remainder of this section we shall concentrate on this frame only. With the aid of the Appendix (A2), we remember that during the irradiation the individual magnetization components will precess in the rotating frame around the effective field $\mathbf{B}_{\text{eff}}$, which is the vector resultant of the resonance offset $b(r)\cdot\mathbf{k}$ and the r.f. magnetic field $B_1\cdot\mathbf{i}$, where $b(r) = (\mathbf{G}_r\cdot\mathbf{r})$. When the irradiation is in the form of intense square pulses, i.e. $B_1 \gg b(r)$ for all r, $\mathbf{B}_{\text{eff}}(r) = \mathbf{B}_1$ for all r. As a result all the components $\mathbf{m}(r)$ of the total magnetization $\mathbf{M}$ will rotate in the zy plane of the rotating frame and on termination of a 90° pulse will all lie parallel to the $-y$ axis. Thus for all r we have $m_y(r) = -m_0(r), m_x(r) = 0, m_z(r) = 0$ and the phase angle $\phi(r)$ of the resultant component xy magnetization $m_{xy}(r)$ relative to the y axis is zero. The net result is therefore that all nuclei within the sample are equally excited.

If the intensity of the irradiation is reduced so that $0 < B_1 \ll b(r)_{\text{max}}$, then the effective field, $\mathbf{B}_{\text{eff}}(r)$, is no longer the same for all values of r. The magnetization of those nuclei in a very weak gradient near $r = 0$ will behave in the same way as for intense pulses, because the condition $\mathbf{B}_{\text{eff}}(r) = \mathbf{B}_1$ is still valid. At the other extreme, where the gradient is large and $b(r) \gg B_1$, $\mathbf{B}_{\text{eff}}$ is effectively along the z axis. Hence for nuclei in these regions, there will be very little movement in the magnetization away from the z axis and so at the end of the pulse $m_x(r) \sim 0, m_y(r) \sim 0$ and $m_z(r) \sim m_0(r)$. It appears therefore that a weak $\mathbf{B}_1$ is a necessary condition for selective excitation of a narrow slice of nuclei at around $b(r) = 0$. This is not however the only condition which has to be met. The boundaries of the excited region must be as sharp as possible and the relative phase of adjacent elemental magnetization vectors must not prevent the observation of a signal.

In the boundary region where $B_1 \sim b(r)$ and $\mathbf{B}_{\text{eff}}$ is parallel neither to the x axis nor to the z axis, all the cartesian components of the elemental magnetization vectors will in general be finite at the end of the square pulse (11). Figure 4.10 illustrates the variation of $m_x(r), m_y(r), m_{xy}(r)$ and $\phi(r)$ with the offset $b(r)$, immediately the pulse has ended. Two points to notice from figure 4.10 are first, that the initial phase of the elemental $m_{xy}(r)$ is distributed over 2π and so the resultant signal will be very small after the termination of the pulse, and secondly, that the boundary of the excited zone of nuclei is not sharp but oscillatory. It will be noted from figure 4.10 that the variation with offset frequency of both $m_y(r)$ and $m_x(r)$ bears a very close resemblance to the real and imaginary parts of the Fourier transform or frequency spectrum of the stimulating pulse. In fact such a resemblance is quite general and as one might therefore suppose, if one chooses to excite the nuclei with a time-domain waveform whose frequency spectrum has a sharp cut-off,

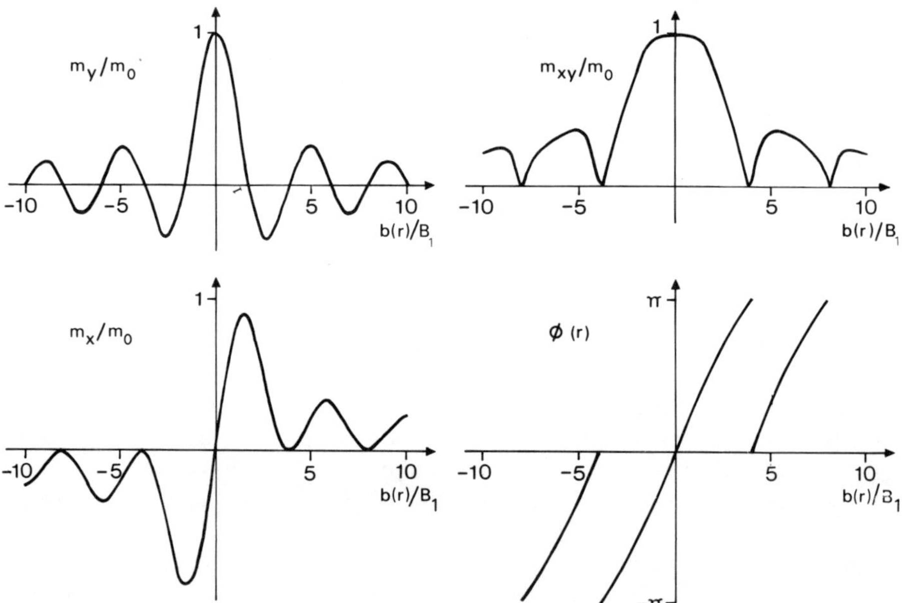

Figure 4.10 A diagram showing the variation with gradient offset $b(r)$, of the rotating frame components of magnetization m_x, m_y and m_{xy}, at the end of a rectangular pulse of r.f. irradiation. The phase angle $\phi(r)$ represents the angle between the magnetization vector m_{xy} and the y axis of the rotating frame when the r.f. pulse terminates.

then the excited slice of nuclei will also have a sharp cut-off. One time-domain envelope which satisfies such a criterion is a *sinc* function (4) and if for the practical expedient of suppressing the wings of this function it is further multiplied by a Gaussian envelope, then the frequency domain spectrum will be the convolution of a rectangle with a narrower Gaussian function, the consequence of which is to endow a slight slope to the sides of the rectangle (12). Excitation with such a modulated r.f. waveform does indeed produce a well defined slice whose edges have a finite width, corresponding to the slight slope of the frequency spectrum.

Having shaped the pulse envelope so as to ensure sharp edges to the slice, it then becomes necessary to overcome the phase distribution which exists at the termination of the pulse, for the elemental magnetization vectors in the xy plane. Fortunately this phase angle is a fairly linear function of the frequency offset. In fact for a rectangular pulse (11), it is almost equal in magnitude to the reduced offset $b(r)/B_1$. As a result of this approximate linearity, a reversal of the gradient at the end of the pulse will cause the elemental magnetization vectors to rephase, such that a refocused signal will occur at some later time, which is in turn dependent on the gradient strength and the intensity of the preceding pulse. The refocused signal arises solely from the irradiated slice and it is therefore possible to observe the nuclear spins in that prechosen slice by making use of switched linear field gradients and selective irradiation in the form of weak, carefully amplitude modulated r.f. waveforms.

4.3.3 *Reconstruction from projections*

A common method for producing an image using NMR is to reconstruct it from a series of projections. This method is very similar to that used in X-ray CT systems etc. One fundamental difference between the various modalities which employ reconstruction from projections is their method for generating the original projections.

By means of a simple example using two elemental sample volumes, we explained briefly in section 4.3.1, how the NMR signal, i.e. the FID, obtained under the influence of a linear magnetic field gradient, provided after Fourier transformation a projection of the nuclear spin density on to the gradient direction. In other words, it provides a measure of the nuclear density as a function of frequency, which because of the linearity of the gradient, corresponds to a scale of position along the gradient direction. Figure 4.11 illustrates three such projections obtained from an array of eight tubes of water. In obtaining figure 4.11 we have also incorporated a slice defining procedure, so that the signal obtained is not from the whole of the object but from a slice through that object. The static gradient which gives rise to

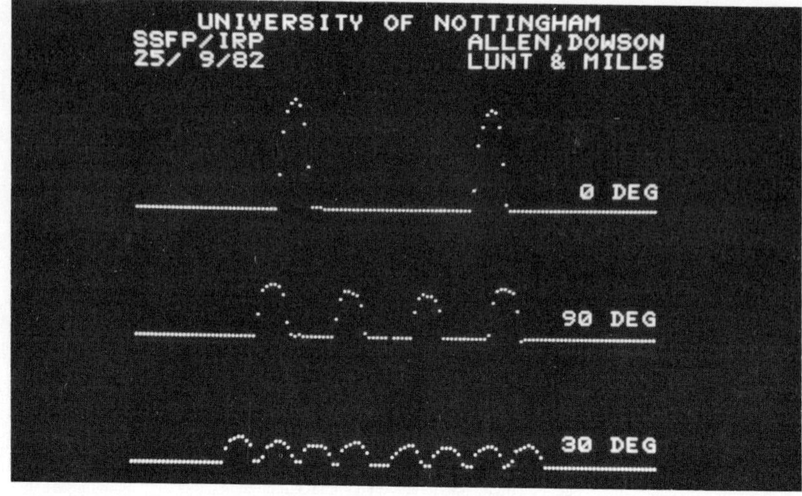

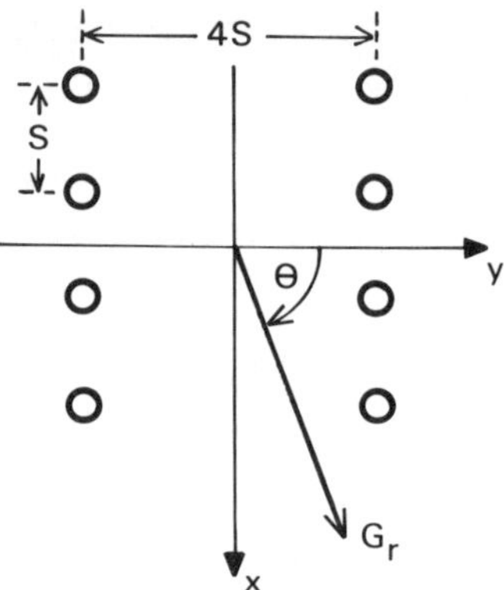

Figure 4.11 A photograph of the CRT display of three projections of a set of eight tubes. The projection angle is measured with respect to *y* axis. At zero degrees, four tubes contribute to each peak. At 90 degrees, each peak contains the signal from two tubes. The signal from each tube is visible when the gradient angle is 30 degrees.

the projection is parallel to the plane of the slice and may be switched on after slice definition has been completed.

In order to reconstruct an image, 128 or 256 projections of a slice are typically used, spanning an angular range of 180° in discrete equal steps. The mathematical details of the procedures for reconstructing an image from these projections were well developed before the advent of NMR imaging (13, 14) and a full account of these procedures can be found in Volume 1 of this series (15). In Volume 1, the emphasis is on the application of reconstruction techniques to tomography with ionizing radiations, e.g. X-rays, where the experimental data comprise one-dimensional spatial projections of an attenuation coefficient and from which a two-dimensional spatial image is reconstructed. By means of the same reconstruction algorithms, spatial NMR images can be obtained using the nuclear-density frequency spectra as the primary projection data. However in NMR, the experimental data (FID) and the final image bear a more convenient relationship than is the case with conventional tomographic experiments and this enables reconstruction to be carried out more simply in NMR imaging, if this is desired.

In common with other imaging modalities, the image in the NMR case is a two-dimensional spatial map which, because spatial displacement and frequency shift are directly proportional in a linear field gradient, is essentially an image in the frequency domain. The experimental FID data however are obtained in the time domain and from the point of view of the two-dimensional Fourier reconstruction technique it is significant that this time-domain FID data is the Fourier transform of the projection data in the frequency domain. In the NMR case therefore, the experimental data and the final image are in Fourier conjugate domains. This is in contrast to most other modalities using Fourier reconstruction techniques, where the experimental data and the final image are both in the spatial domain and the Fourier conjugate domain acts only as a means to an end.

To illustrate this point very briefly, we recall that corresponding to a two-dimensional spatial density-function $F(x, y)$, which we call the image, there must exist a two-dimensional Fourier transform of this function i.e. $\phi(\alpha, \beta)$ in a Fourier conjugate space, such that

$$F(x, y) = \iint_{-\infty}^{\infty} \phi(\alpha, \beta) \exp\{-i2\pi(x\alpha + y\beta)\}\, d\alpha\, d\beta \tag{4.11}$$

In order to employ the projection theorem (13), this equation is subjected to a co-ordinate transformation which converts the cartesian variables x and y into cylindrical polar radial and angular variables r and a respectively, and similarly converts α and β into cylindrical polar ρ and θ, subject to the special condition that a and θ are equal. We then have the corresponding equation

$$f(r, \theta) = \int_0^{\pi} \int_{-\infty}^{\infty} \chi(\rho, \theta) \exp\{-i2\pi r\rho\} |\rho|\, d\rho\, d\theta, \tag{4.12}$$

where the $|\rho|$ arises during the transformation of the differential coefficient $d\alpha d\beta$. If we now identify r, not with spatial displacement but with its proportional frequency shift, v, in a longitudinal gradient along a line at an angle θ and correspondingly identify ρ with the time variable, then the inner integral of equation (4.12) may be rewritten

$$\int_{-\infty}^{\infty} M_{\theta}(t) \exp\{-i2\pi\nu t\} |t| \, dt$$

where M is written for χ to emphasize that it is the magnetization which is the function of time. Now this integral corresponds to a filtered projection, where the filter function is simply $|t|$. Thus in order to obtain the filtered projection, we multiply the FID by the time variable before performing the Fourier transform. To accommodate the outer integral and at the same time generate a density image on a cartesian matrix, a series of such filtered profiles, each one corresponding to a different value of θ, is back-projected and interpolated on to that cartesian matrix in a manner identical to that used in other CT modalities.

In order to obtain a three-dimensional, 3D, NMR image, rather than a two-dimensional, 2D, slice, the original data set of fixed gradient FIDs is now a sequence in which the field gradient direction is stepped through a mesh of points distributed over the surface of a sphere. From such a data set 3D images have been generated (16) using

(i) a two-stage algorithm, which involves two successive computational stages of 2D reconstruction, each using the same algorithm and employing an intermediate data set of 2D projections,
(ii) a direct algorithm, which generates a complete 3D image in one computational process using the inversion formula of the 3D Radon transform, and
(iii) a hybrid algorithm (17), which merges the essential features of (i) and (ii), employing 3D type filtering in conjunction with two-stage back projection.

Although the analytical schemes of the two-stage and the direct algorithms are equivalent, the two-stage method places constraints on the choice of gradient direction but has the advantage of requiring a shorter computational time for a complete 3D image. This is because the two-stage method is structured to take advantage of the special nature of its angular grid. The direct method does not impose conditions on the gradient directions and although it involves more computational steps for a full 3D image, becomes comparable in speed or even faster than the two-stage method when the extent of the image is restricted to a plane or a small region of pixels. Moreover, the direct method displays a freedom from peripheral artefacts not shown by the two-stage method.

When one employs projections in order to reconstruct an image, the FID's which one uses to generate those projections need not necessarily be the nuclear response to a single 90°pulse. It is necessary only that the dephasing of the magnetization components be governed by the linear gradients alone. As a result the nuclear signal may indeed be a straightforward single pulse FID or it may be an echo generated by several pulses or it may be the decay of the refocused signal following slice-defining irradiation.

4.3.4 *Fourier zeugmatography*

This is a technique which owes its introduction to Kumar, Welti and Ernst (18) and which employs a two-dimensional Fourier transform of the NMR data. It must not be confused with two-dimensional Fourier reconstruction from projections as outlined in section 4.3.3, because it does not employ a sequence of projections each obtained under the influence of a single fixed gradient. Instead it employs two orthogonal linear gradients sequentially, as is illustrated in figure 4.12. The purpose of the latter gradient, G_y, is to encode one spatial cartesian co-ordinate, namely the y co-ordinate, into the NMR signal, by governing the elemental magnetization rotation frequencies in the rotating frame and hence governing the oscillations of the component EMF during the data sampling time t_y. This part is entirely analogous to the acquisition of projection data as described earlier in section 4.3.1. The purpose of the former gradient, G_x, is to encode the orthogonal cartesian co-ordinate, x, into the NMR signal, by governing the phase of $\mathbf{M}_{xy}$ at the commencement of the latter gradient evolution period i.e. at $t = t_x$. This phase angle, ϕ, depends on both the magnitude of G_x and its time duration t_x. In order to generate the complete NMR data set, sequences of FIDs are sampled in t_y, each corresponding to a different value of ϕ and to bring about the variation in ϕ, one may step either G_x or t_x between FIDs.

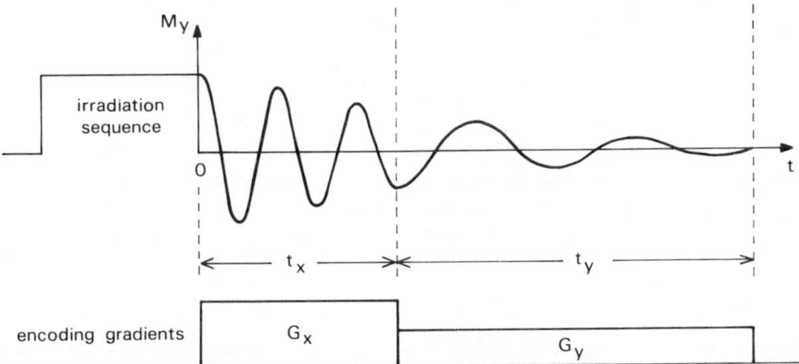

Figure 4.12 A diagram illustrating the sequence of gradients used in Fourier zeugmatography. The signal from a magnetization element will display a different frequency in each gradient period. The signal is collected in the period t_y, but the starting phase of this signal is determined by the oscillations in t_x.

To illustrate the treatment of the data, consider two elemental samples placed asymmetrically in the xy plane at locations ξ and ζ. This allows us to concentrate on the evolution of the elemental magnetizations and to assume that slice definition procedures have been taken care of in the r.f. irradiation sequence. As the magnetization evolves after the irradiation but under the influence of G_x, the sample at ξ will contribute a magnetization component precessing about z at a frequency $\Omega_\xi = G_x \cdot x_\xi$ in the rotating frame. If this magnetization component has $\phi_\xi(0) = 0$ at $t = 0$, then at $t = t_x$

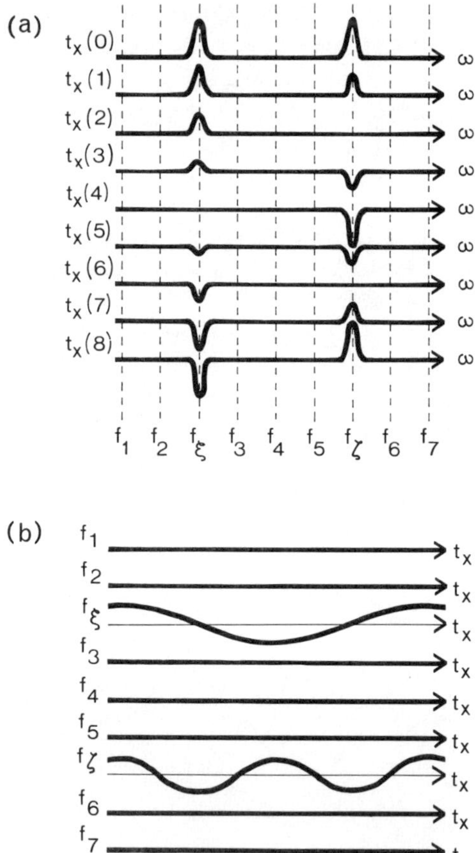

Figure 4.13 An illustration of part of the two-dimensional data arrays used in Fourier zeugmatography. Part (*a*) shows a section of the stack of the frequency domain data following the first Fourier transform of the time domain data. The labels $t_x(i)$ indicate the length of the pre-conditioning period in the x gradient prior to data sampling in t_y. Part (*b*) is a transposition of the data in part (*a*), such that data points down the vertical dotted lines of part (*a*) are displayed horizontally as a function of t_x.

we have

$$\phi_\xi(t_x) = \Omega_\xi \cdot t_x = G_x \cdot x_\xi \cdot t_x \tag{4.13}$$

This means that the period of influence of G_y will commence with

$$\left.\begin{aligned} m_{y\xi} &= m_{0\xi} \cos \Omega_\xi t_x \\ m_{x\xi} &= m_{0\xi} \sin \Omega_\xi t_x \end{aligned}\right\} \tag{4.14}$$

in the rotating frame. The evolution during t_y will impose a different precessional frequency, $f_\xi = G_y \cdot y_\xi$, on the elemental magnetization from the sample at ξ and the rotating-frame components then become

$$\left.\begin{aligned} m_{y\xi} &= m_{0\xi} \cos \{\Omega_\xi t_x + f_\xi(t - t_x)\} \\ m_{x\xi} &= m_{0\xi} \sin \{\Omega_\xi t_x + f_\xi(t - t_x)\}. \end{aligned}\right\} \tag{4.15}$$

Similar expressions for the rotating-frame magnetization components of the sample placed at ζ, can be obtained by exchanging the frequencies Ω_ξ and f_ξ for Ω_ζ and f_ζ respectively in equation (4.15). During an experiment, the nuclear signal is collected during the time interval t_y and quadrature phase-sensitive detection of this signal at the central resonant frequency $\omega_0 = \gamma B_0$, will produce two FID waveforms, corresponding to the y component sum and the x component sum of the ξ and ζ elemental magnetization as given by equation (4.15).

Complex Fourier transform of the two FID waveforms with respect to

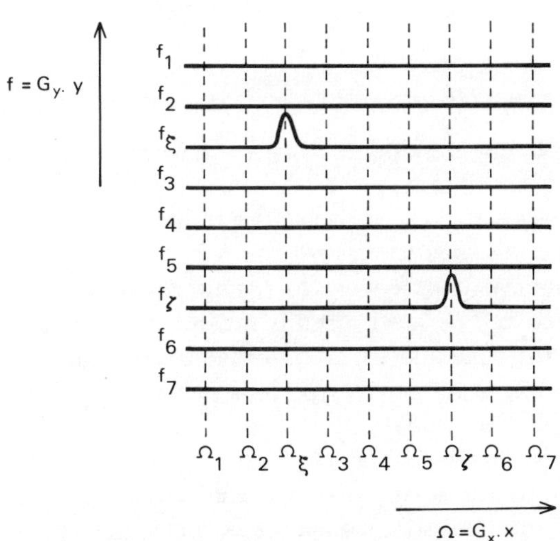

Figure 4.14 A diagram showing part of the two-dimensional data array following the second Fourier transform in a Fourier zeugmatography experiment. The frequency axes f and Ω are directly proportional to the space co-ordinates y and x respectively.

$(t-t_x)$ will produce frequency domain components at both f_ξ and f_ζ. However, the relative magnitudes of the real and imaginary parts of these components will be governed by the phases ϕ_ξ and ϕ_ζ at time $(t-t_x)=0$. Figure 4.13 illustrates part of a sequence of such frequency-domain envelopes, obtained from FID waveforms initiated after different values of t_x. From this figure, which does not show the imaginary parts, it is clear that the magnitude of the real part oscillates in t_x and the oscillation frequencies are governed by G_x, x_ξ and x_ζ. Suppose we imagine that a complete sequence of frequency profiles, stacked as in figure 4.13(*a*) and constituting a two-dimensional data matrix, is now transposed in a manner similar to that shown between figures 4.13(*a*) and (*b*), where the time axis becomes horizontal and the frequency, f_i, now labels each time domain waveform. A second Fourier transform, this time with respect to t_x, will give rise to a two-dimensional frequency map of nuclear density as shown in figure 4.14. In figure 4.14, the frequency axes can be replaced by space coordinate axes, because of the linearity of the field gradients and by converting the Fourier amplitudes into pixel intensity this map becomes a 2D NMR image.

When exploiting the technique of Fourier zeugmatography, the nuclear signal may arise in a number of ways. For example, it may be the response to a single pulse, it may be an echo stimulated by several pulses or it may be the refocused signal following slice-defining irradiation. In this respect Fourier zeugmatography is like the use of projections, in that it is a technique for generating a 2D image and at the same time leaves much flexibility in the method of stimulation of the primary signal.

4.3.5 *Echo-planar imaging*

Techniques which make use of reconstruction from projections or of Fourier zeugmatography to obtain a two-dimensional image, always employ a sequence of nuclear signals, between the individual members of which, a recovery period is allowed to elapse. The purpose of this recovery period is to enable the nuclear spin system to regain its thermal equilibrium state. However, the recovery period is usually much longer than the irradiation or the signal sampling time and hence it dominates the data collection time. In order to avoid successive recovery periods and hence to provide a very fast imaging technique, Mansfield and coworkers (19, 20) have devised what is known as the *echo-planar* technique. The echo-planar technique attempts to retrieve the information on intensity encoded with orthogonal spatial co-ordinates simultaneously, from a single nuclear signal. It does this by stimulating repetitive echoes which keep alive the nuclear signal for a relatively long period of time.

The outline of an echo-planar sequence is illustrated in figure 4.15.

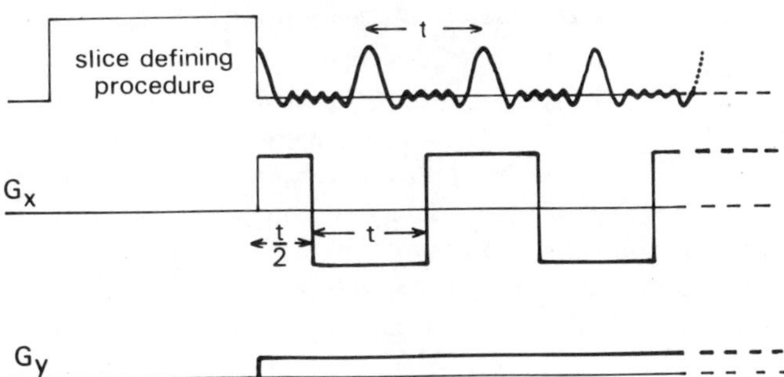

Figure 4.15 An outline of the gradient waveforms used in Echo Planar imaging. The G_z gradient together with selective excitation are incorporated into the slice-defining procedure, whereas the x and y gradients are used to encode the NMR signal with spatial information. Reversal of the larger gradient G_x results in the refocusing of an echo signal during each gradient half cycle.

Following slice definition, the nuclear magnetization is allowed to evolve under the influence of two orthogonal gradients G_x and G_y. G_x is periodically reversed and therefore brings about a refocused echo signal every half cycle. If relaxation effects could be ignored and if G_y were zero, then the precession of each elemental magnetization component would be exactly reversed by the reversal of the G_x gradient and echoes would continue to reappear indefinitely. Moreover, the shape of each echo, when Fourier transformed would give the nuclear density projected onto the x direction. If the y gradient were now made finite, but small in comparison with G_x, then although refocusing would take place, it would be by no means perfect. For example, at each location in space the gradient offset fields due to G_x and G_y would add in one half cycle but subtract in other. Therefore precession of the elemental magnetizations in one direction will not exactly balance their precession in the opposite direction. As a result a phase mismatch builds up with each successive cycle. The rate with which this phase mismatch builds up is governed by G_y and the y displacement. Thus the overall decay of the echo train gives rise to spatial information in the y direction.

Because spatial information is contained not only within the shape of one echo, but also in overall decay of the echo train, the nuclear signal is sampled over a long series of echoes. From a single Fourier transform of the echo train all the spatial information is obtained. However, two points are crucial to the viability of the method. The first is the repetitive nature of the echoes and the second is the relative amplitudes of the gradients. Very roughly, the periodic nature of the echoes can be represented by the convolution of one echo with a comb of delta functions in the time domain (4). When Fourier transformed, this convoluted waveform gives rise to the transform of a single echo, multiplied by a comb of delta functions in the frequency domain, namely a set of spikes whose amplitudes conform to the envelope of the Fourier transform of the single echo. In other words, the repetitive nature of the echo train endows a discreteness to the x projection of the nuclear density as shown in figure 4.16 and the repetition frequency determines the interval between the discrete spikes. The decay of the echo train due to the much weaker y gradient will lead to a broadening of the spikes. In fact, the details of the decay, and therefore of the broadening, will be determined by how the nuclei are distributed in the y gradient. Thus if G_y is small enough not to cause the broadening to exceed the discrete spacing of the spikes, then both the x and y coordinates can be obtained from a single Fourier transform. In figure 4.16, because of the hole in the middle of the annulus, the projection spikes of the central region are split as well as broadened.

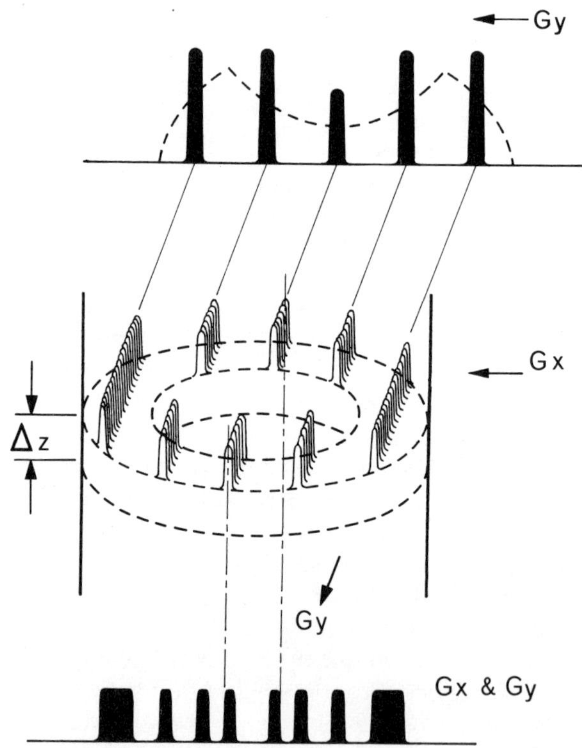

Figure 4.16 An illustration of the projections obtainable from a homogeneous annulus. The upper dotted projection is that which would be obtained in a static single gradient G_x, whereas if G_x is periodically reversed this turns into the discrete profile which is superimposed. The lower projection corresponds to that obtained when a small continuous G_y is added to the periodically reversed G_x. P. Mansfield and I. L. Pykett, *J. Mag. Res.* **29** (1978) 355.

Using this technique Mansfield has obtained images in fractions of a second, as compared with a few minutes for the more conventional techniques. At the moment this speed is being obtained at the expense of picture quality. Nevertheless, if real-time moving pictures are required then a lower quality of image can be tolerated because the eye does its own measure of averaging.

4.4 Discrimination using relaxation rates

Section 4.3 described the techniques commonly employed to obtain an NMR image. During this description little attention was paid to the effects of relaxation and in the interests of clarity, the image was taken to correspond to the nuclear density alone. Such an assumption is not entirely valid because the relaxation processes described in section 4.2.2 can have a marked effect on the magnitude of the magnetization and hence on the intensity of the image. This dependence of image intensity on the relaxation rates presents us with the possibility of using these rates as a discrimination mechanism, provided that there is a meaningful variation of relaxation rates among the biological tissues. Such a variation does in fact exist and is sufficient for relaxation discrimination to have become a powerful extension of NMR imaging.

4.4.1 *Relaxation in biological systems*

The variation in the proton relaxation rates between the tissues of the major organs depends largely on their different water content. This conclusion has been drawn from a substantial amount of data on excised tissue from mice, rats, rabbits and human specimens, which show trends which display a reassuring degree of independence from the strain of the animal and also from the animal type. More thorough reviews of this relaxation data have been made by Mansfield and Morris (2) and by Taylor and Bore (21). In summary, tissue types can be roughly characterized in terms of their water contents. For instance in mice, liver occurs at the lower end of the range, containing between 66% and 70% water. The brain appears at the upper end with a water concentration approaching 80%. Figure 4.17 illustrates by means of a bar chart the variation in water content for the tissues of mice between the liver and the brain and in addition shows the corresponding trend in T_1. A similar trend is followed by T_2, though the absolute values are approximately an order of magnitude less than those for T_1. A significant feature of this trend and one which makes NMR very sensitive to water concentration is that a 15% to 20% change in water concentration produces a change in relaxation rate of about 200%.

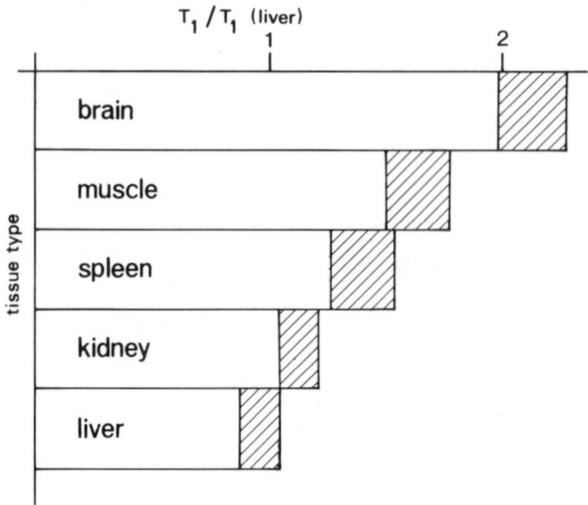

Figure 4.17 An indication of the variation of T_1 between some of the major organs in mice. The absolute values of T_1 depend on the r.f. and on the temperature.

The T_1 dependence on the change in water concentration, ρ, can be explained in terms of the exchange of water molecules within the cellular system. The exchange takes place between those sites closely linked to the macromolecules and those sites well separated from the macromolecules and which therefore resemble bulk water. Experiments on samples containing their water in a single compartment e.g. largely intracellular water in barnacle muscle (22), showed a magnetization recovery along the z direction which was a single exponential. For this reason it is usually assumed that rapid exchange takes place between the two types of water site, thus giving rise (23) to a single observed relaxation rate R_{obs}, which is the population weighted average of the rates R_m and R_b corresponding to the macromolecular linked and the bulk water sites respectively, i.e.

$$R_{obs} = R_m \cdot p_m + R_b \cdot (1 - p_m) \tag{4.16}$$

where p_m is the relative population of the water molecules associated with the macromolecules. Hence

$$R_{obs} = R_b + (R_m - R_b) \cdot p_m \tag{4.17}$$

and p_m is in turn directly related to the fractional water concentration ρ in the tissue. The slower motions associated with the macromolecules, e.g. internally rotating groups or overall tumbling and translation, will cause those water molecules associated with them to have the more efficient spin-lattice relaxation, i.e. $R_m > R_b$ and therefore $(R_m - R_b)$ is positive. By increasing the water concentration, one increases the number of bulk water molecules and hence decreases the proportion p_m, of efficient relaxers. In consequence, this will make the overall average relaxation less efficient, i.e. makes R_{obs} smaller and hence increases the resulting T_1.

Although the body of experimental data, which measures T_1 and ρ from a variety of excised tissues, exhibits a significant degree of scatter, it is nevertheless consistent with a fairly linear dependence of T_1 on ρ. The enhanced sensitivity of T_1 measurements arises because the slope of this linear variation is greater than unity. In fact, a slope of

$$\frac{dT_1}{d\rho} \sim 8$$

can be obtained from the accumulated data (2).

T_2 is similarly affected by the exchange between the two types of site. In the rapidly mobile environment of bulk water, the local magnetic fields will be well averaged out and the dephasing of the nuclear magnetization will be slow. Macromolecules on the other hand will contain many fixed protons which restrict the degree of local field averaging for water protons in their vicinity. The dephasing of water proton magnetization in these locations will therefore be more rapid. Exchange between the two types of site, if fast on the time scale of the dephasing, will tend to average the local field dephasing to some value intermediate between the two extremes.

Biological systems in general are complicated by the fact that the water within them resides in compartments, between which the passage of protons (or spin information by spin diffusion) may be seriously restricted, e.g. as with extra- and intracellular water. If, moreover, the relaxation behaviour in the separate compartments is not the same, then the overall observed relaxation will no longer be describable in terms of a single exponential. One must therefore be prepared first to recognize the fact that the relaxation in biological systems may exhibit more than one component exponential, and second, to live with the notion of a representative relaxation rate for a particular tissue type.

4.4.2 *Some methods for encoding images with relaxation information*

In the interests of speed, instead of attempting to make a quantitative measurement of T_1 or T_2 at each pixel location of a 2D image, what is often done is to modulate the individual pixel intensities by the relaxation time of interest. This can be brought about quite easily by allowing the relaxation rate in question to govern the magnitude of the magnetization which induces the nuclear signal, instead of allowing the magnetization to recover to its full thermal equilibrium value before each projection, etc., is taken. There

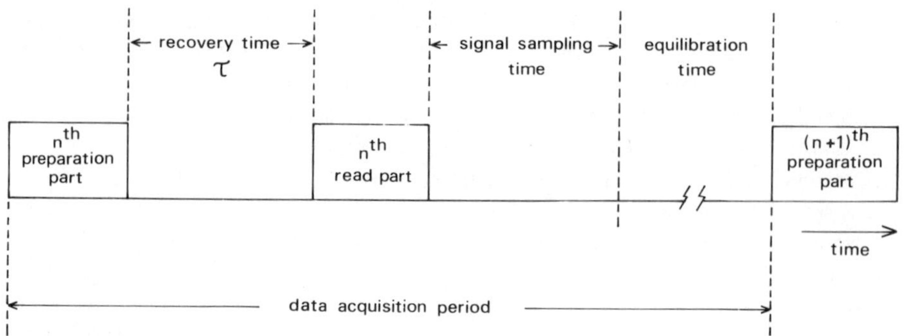

Figure 4.18 A schematic diagram of a single signal acquisition period, a long series of which will constitute a signal averaging sequence. The preparation part, together with the recovery time τ, determines the proportion of the total magnetization which will be effective during the signal sampling time. The r.f. irradiation and gradient waveforms which form the read part serve to encode the nuclear signal with spatial information.

are a number of variations on this theme and in the following subsections those denoted by the terms *saturation recovery*, *inversion recovery* and *spin-echo imaging*, will be described. Before describing their individual differences, it is useful to recall the general strategy behind all of these methods. Because of the inherent weakness of the nuclear signal, signal averaging is always necessary in NMR imaging. As a result the acquisition of identical signals is repeated many times. Figure 4.18 illustrates a single signal acquisition period, which when repeated will give rise to a much longer signal averaging sequence. In figure 4.18 we have subdivided the irradiation into a *read part*, which stimulates the space encoded signal for imaging purposes and a *preparation part* which preconditions the magnetization and together with the *recovery time*, τ, controls the magnitude of the magnetization effective in the read part. If the image were only required to display proton density, then the preparation part would be absent, τ would be zero and the equilibration time would be sufficient to allow complete recovery to thermal equilibrium before the next acquisition commenced. If, instead, the image is required to display some relaxation information in order to enhance the discrimination, then one or other of the following schemes may be employed.

Saturation recovery. This technique can be used to discriminate against regions of long T_1. For example, suppose the preparation part of figure 4.18 were again omitted, that the read part produced a 90° rotation of the z magnetization and that the equilibrium time were shortened to prevent full recovery to thermal equilibrium. In effect the nth read part would now be acting as the $(n + 1)$th preparation part and the sum of the signal sampling and equilibration times is effectively a recovery time. Following the irradiation of a read part, the z magnetization will recover according to the curve marked $p=1$ in figure 4.6 and the shorter the T_1, the more magnetization will have recovered before the next irradiation. Thus in the sum of many data acquisition periods, the resultant signal from regions of longer T_1 will be depressed relative to that from shorter T_1 regions and the short T_1 regions will therefore be highlighted.

The ability of the saturation recovery technique to depress signal from regions whose T_1 is longer than the data acquisition period can be exploited to display fluid moving through the slice being imaged. By making the data acquisition period quite short relative to T_1 of biological water, most of the nuclei which reside permanently in the imaged slice will become saturated and give rise to very little signal. In contrast the fluid which is flowing into the slice will be fresh from an NMR point of view, because it will have escaped magnetization flipping by irradiation bursts prior to its arrival in the slice. The fluid therefore exhibits a large magnetization and gives rise to a bright region in the image.

Inversion recovery. Discrimination against a band of T_1 values can be obtained with this technique and moreover, the band itself can be adjusted to coincide with either the shorter or the longer end of the T_1 range of biological water. To perform inversion recovery, the preparation part of the irradiation pattern needs to cause a 180° inversion of the z magnetization and this can be brought about by either a 180° pulse or an adiabatic fast passage (24). Following the 180° inversion, the z magnetization will recover according to the curve marked $p=2$ in figure 4.6 and the crux of discrimination by inversion recovery is to choose the recovery time, τ, to be equal to the average t_{NULL} for the band of T_1 values against which the discrimination is required. By moving τ around, discrimination against different T_1 bands can be accomplished. Once τ has been chosen to depress the signal from a particular band of T_1, one must bear in mind that the regions with a longer T_1 will give rise to a negative magnetization and those with much shorter T_1s will have a positive magnetization.

The inversion recovery technique is also being used (25) to obtain a quantitative estimate of the variation in T_1 values within an imaged slice. The validity of this estimate rests on the assumption that the recovery of the z magnetization can be described everywhere by a single exponential, though that exponential may itself vary from place to place within the slice. If single exponential recovery is indeed true, then measurements of the magnitude of the magnetization at two different times during the recovery will be sufficient to determine T_1. This means that for each projection, etc., two sets of data need to be obtained corresponding to the two different recovery times. By defining a pixel intensity or colour scale in terms of the T_1 estimates alone, it is then possible to display a T_1 image.

Spin-echo imaging. In order to exploit the spin–spin relaxation time, T_2, as a discrimination parameter one employs spin-echoes. The preparation part of the irradiation sequence must then be such as to generate these echoes, possibly many of them, and figure 4.19 illustrates the basic sequence required. By irradiating initially with a 90° pulse to rotate the magnetization to the xy plane, one can refocus successive echoes by periodically flipping the xy magnetization through 180° by means of a 180° r.f. pulse (26). If the gradients remain constant and no spin–spin relaxation takes place, then the 180° flipping of the xy magnetization distribution should lead to perfect refocusing. However, spin–spin relaxation does take place, therefore refocusing is not perfect and successive echoes progressively decrease in size. For a single component system this decrease will progress with the characteristic time constant T_2. Within the heterogeneous slice producing the image there will be many component T_2s and the contribution to the echoes from those regions with a short T_2 will decrease faster than the contributions from regions with a long T_2. Thus, by using an echo well separated from the initial 90° pulse, those contributions from short T_2 regions will be depressed and long T_2 regions will be highlighted. Comparative studies can be made by constructing a series of images, each one resulting from the corresponding echo in each echo train.

It is interesting to note that in T_1 modulated images, particularly those resulting from saturation recovery, it is the *short* T_1 region which shows up brightly, whereas with spin-echo imaging the *long* T_2 regions are most intense. However, in liquid-like systems T_1 and T_2 go hand in hand. When one is long, so is the other and vice versa. This means that complementary images can be obtained by modulating with T_1 on the one hand or with T_2 on the other.

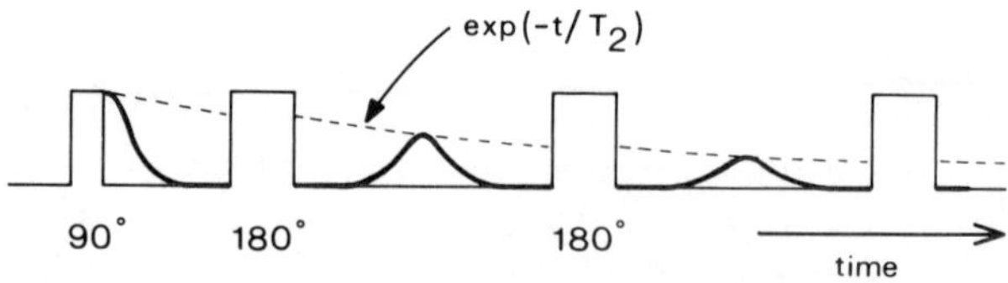

Figure 4.19 A simple spin-echo generating sequence, where the 90° pulse turns the z magnetization into the xy plane and following the dephasing of the subsequent signal in a field gradient, successive 180° pulses turn over the dephasing pancake of spins to ensure refocusing in the same gradient. The separation of the 180° pulses is twice the interval between the 90° pulse and the first 180° pulse and the echoes occur mid-way between the 180° pulses.

Appendix

A.1 The dipole–dipole interaction

The classical interaction energy between two magnetic dipoles $\boldsymbol{\mu}_1$ and $\boldsymbol{\mu}_2$ separated by the vector displacement $\mathbf{r}_{12}$ is

$$\frac{\boldsymbol{\mu}_1 \cdot \boldsymbol{\mu}_2}{r_{12}^3} - \frac{3(\boldsymbol{\mu}_1 \cdot \mathbf{r}_{12})(\boldsymbol{\mu}_2 \cdot \mathbf{r}_{12})}{r_{12}^5}$$

By substituting quantum mechanical spin operators for the classical magnetic moment vectors, the interaction Hamiltonian $\mathscr{H}_{\mathrm{Dip}}$ is obtained as

$$\mathscr{H}_{\mathrm{Dip}} = \frac{\gamma^2 \hbar^2}{r_{12}^3} \{\mathbf{I}_1 \cdot \mathbf{I}_2 - 3(\mathbf{I}_1 \cdot \mathbf{r}_{12})(\mathbf{I}_2 \cdot \mathbf{r}_{12})/r_{12}^2\} \tag{A1.1}$$

In order to make clear the various functions of this Hamiltonian, it is useful to rewrite it in terms of the z component operator I_z, together with the raising and lowering operators I_+ and I_- ($I_\pm = I_x \pm iI_y$). Doing this we obtain (27)

$$\mathscr{H}_{\mathrm{dip}} = \frac{\gamma^2 \hbar^2}{r_{12}^3} \{A + B + C + D + E + F\}$$

where

$$\begin{aligned}
A &= (1 - 3\cos^2 \theta_{12}) I_{1z} I_{2z} \\
B &= -\tfrac{1}{4}(1 - 3\cos^2 \theta_{12})\{I_{1+} \cdot I_{2-} + I_{1-} \cdot I_{2+}\} \\
C &= -\tfrac{3}{2}\{\sin \theta_{12} \cdot \cos \theta_{12} \cdot \exp(-i\phi_{12})\}\{I_{1z} I_{2+} + I_{1+} I_{2z}\} \\
D &= C^* \\
E &= -\tfrac{3}{4}\{\sin^2 \theta_{12} \exp(-2i\phi_{12})\} I_{1+} I_{2+} \\
F &= E^*
\end{aligned}$$

and r_{12}, θ_{12} and ϕ_{12} are the polar co-ordinates describing the relative positions of the two spins in a frame of reference whose z axis is defined by the applied magnetic field $\mathbf{B}_0$.

When written in this form, it is clear that each term in dipolar Hamiltonian is the product of a space part and a spin operator. The space part of the dipolar interaction will be modulated by molecular motion, which causes the space co-ordinates to become time-dependent. To see the effect of the spin operators, one must remember that NMR is usually performed in a large static magnetic field $\mathbf{B}_0$ and that the dipolar interaction is very small compared with the Zeeman interaction in $\mathbf{B}_0$. In the Zeeman representation where $|\alpha\rangle$ represents the spin parallel eigenstate and $|\beta\rangle$ the antiparallel eigenstate, we have

$$I_z|\alpha\rangle = \tfrac{1}{2}|\alpha\rangle; \quad I_z|\beta\rangle = -\tfrac{1}{2}|\beta\rangle$$

and

$$I_+|\beta\rangle = |\alpha\rangle; \quad I_-|\alpha\rangle = |\beta\rangle$$

Thus the dipolar term A has the effect of producing energy level shifts and for a system composed of a large number of spins, leads to a broadening of the frequency spectrum. Term B, known as the *flip-flop* term, can facilitate the transfer of spin energy within the spin system itself. It does this by causing two oppositely oriented nuclear spins to reverse their orientations simultaneously. It therefore affects the lifetime of the spin states and thence T_2. Spin-lattice relaxation is brought about by terms C to F, which are able to stimulate net transitions between energy states through the operators I_+ and I_-. The two terms C and D cause transitions at the frequency ω_0, whereas E and F generate a stimultaneous transition of two spins and therefore cause an exchange of energy with the lattice at an energy of $2\hbar\omega_0$. Both T_1 and T_2 can therefore be affected by the dipolar interaction between two neighbouring nuclear spins.

A.2 The Bloch equation

The objective of this section is to recast the Bloch equation in matrix form, to transform it to a rotating reference frame thus showing the simplification obtained and to illustrate its solution by solving a particularly simple case.

A.2.1 The Bloch equation in matrix form. Recalling the Bloch equation from section 4.2.3, we have

$$\dot{\mathbf{M}} = \gamma \mathbf{M} \wedge \mathbf{B} - \frac{\{M_x \cdot \mathbf{i} + M_y \cdot \mathbf{j}\}}{T_2} + \frac{\{M_z - M_0\}\mathbf{k}}{T_1}. \tag{A1.2}$$

To simplify the situation we shall consider a single elemental magnetization component and to emphasize this we shall denote the magnetization by **m**. Writing $R_1 = 1/T_1$ and $R_2 = 1/T_2$ equation (A1.2) may be rewritten

$$\begin{aligned}\dot{\mathbf{m}} = \{\gamma(m_y B_z - m_z B_y) - R_2 m_x\}\mathbf{i} + \{\gamma(m_z B_x - m_x B_z) - R_2 m_y\}\mathbf{j} \\ + \{\gamma(m_x B_y - m_y B_x) - R_1 m_z + R_1 m_0\}\mathbf{k}\end{aligned}$$

and if $\mathbf{m} = (0, 0, m_0)$ this may be further simplified to

$$\dot{\mathbf{m}} = \mathbf{Q}\mathbf{m} + R_1 \mathbf{m}_0 \tag{A1.3}$$

where

$$\mathbf{Q} = \begin{bmatrix} -R_2 & \gamma B_z & -\gamma B_y \\ -\gamma B_z & -R_2 & \gamma B_x \\ \gamma B_y & -\gamma B_x & -R_1 \end{bmatrix}.$$

Because the diagonal elements of **Q** are relaxation rates whereas the off-diagonal elements are frequencies it is often convenient to subdivide **Q** as

$$\mathbf{Q} = \mathbf{A} + \mathbf{T}$$

where **A** is the off-diagonal part and **T** contains the diagonal relaxation rates. In general in imaging experiments we may write

$$\mathbf{B} = B_0\mathbf{k} + (\mathbf{G}_r \cdot \mathbf{r})\mathbf{k} + \mathbf{B}_1.$$

If $\mathbf{B}_1$ rotates counterclockwise in the xy plane at a frequency $\omega_0 = \gamma B_0$, then by introducing the notation $\mathbf{G}_r \cdot \mathbf{r} = b$, $\gamma(B_0 + b) = \omega$ and $\gamma B_1 = \omega_1$, the matrices **A** and **T** become

$$\mathbf{A} = \begin{bmatrix} 0 & \omega & \omega_1 \sin\omega_0 t \\ -\omega & 0 & \omega_1 \cos\omega_0 t \\ -\omega_1 \sin\omega_0 t & -\omega_1 \cos\omega_0 t & 0 \end{bmatrix}; \qquad \mathbf{T} = \begin{bmatrix} -R_2 & 0 & 0 \\ 0 & -R_2 & 0 \\ 0 & 0 & -R_1 \end{bmatrix}. \tag{A1.4}$$

A.2.2 Transformation to a rotating reference frame. Suppose the rotating reference frame rotates about the z axis in a negative sense and at a frequency Ω. Denoting the magnetization in the rotating frame by $\mathbf{m}_\rho$, we have

$$\mathbf{m}_\rho = \begin{bmatrix} \cos\Omega t & -\sin\Omega t & 0 \\ \sin\Omega t & \cos\Omega t & 0 \\ 0 & 0 & 1 \end{bmatrix} \begin{bmatrix} m_x \\ m_y \\ m_z \end{bmatrix}. \tag{A1.5}$$

Writing the transformation matrix as $\mathbf{U}^{-1}$ and differentiating with respect to time equation (A1.5) gives

$$\dot{\mathbf{m}}_\rho = \dot{\mathbf{U}}^{-1}\mathbf{m} + \mathbf{U}^{-1}\dot{\mathbf{m}}.$$

Substituting for $\dot{\mathbf{m}}$ from equation (A1.3) yields

$$\dot{\mathbf{m}}_\rho = \dot{\mathbf{U}}^{-1}\mathbf{m} + \mathbf{U}^{-1}\{\mathbf{Q}\mathbf{m} + R_1\mathbf{m}_0\}$$

and substituting further for **m** and remembering that $\mathbf{m}_{0\rho} = \mathbf{m}_0$, $\mathbf{m}_\rho$ becomes

$$\dot{\mathbf{m}}_\rho = \{\dot{\mathbf{U}}^{-1}\mathbf{U} + \mathbf{U}^{-1}\mathbf{Q}\mathbf{U}\}\mathbf{m}_\rho + \mathbf{U}^{-1}R_1\mathbf{m}_{0\rho}.$$

From the definition of $\mathbf{U}^{-1}$ given in equation (A1.5)

$$\dot{\mathbf{U}}^{-1}\mathbf{U} = \begin{bmatrix} 0 & -\Omega & 0 \\ \Omega & 0 & 0 \\ 0 & 0 & 0 \end{bmatrix}.$$

and

$$\mathbf{U}^{-1}\mathbf{QU} = \mathbf{U}^{-1}\mathbf{AU} + \mathbf{U}^{-1}\mathbf{TU} = \begin{bmatrix} 0 & \omega & 0 \\ -\omega & 0 & \omega_1 \\ 0 & -\omega_1 & 0 \end{bmatrix} + \begin{bmatrix} -R_2 & 0 & 0 \\ 0 & -R_2 & 0 \\ 0 & 0 & -R_1 \end{bmatrix}.$$

If the rotating frame rotates at exactly the resonance frequency, i.e. $\Omega = \omega_0$ and writing $(\omega - \omega_0) = \omega_b = \gamma b$ then

$$\dot{\mathbf{m}}_\rho = \mathcal{Q}\mathbf{m}_\rho + R_1 \mathbf{m}_{0\rho} \tag{A1.6}$$

where

$$\mathcal{Q} = \begin{bmatrix} 0 & \omega_b & 0 \\ -\omega_b & 0 & \omega_1 \\ 0 & -\omega_1 & 0 \end{bmatrix} + \begin{bmatrix} -R_2 & 0 & 0 \\ 0 & -R_2 & 0 \\ 0 & 0 & -R_1 \end{bmatrix}. \tag{A1.7}$$

Thus if the rotating frame rotates in the same sense and at the same frequency as the resonant radiation, then the matrix operator $\mathcal{Q}$ is time independent and this greatly simplifies the solution of the differential equation (A1.6).

A.2.3 A simple solution of rotating frame Bloch equation. Restricting the problem to one in the extreme narrowing limit, where $R_1 \sim R_2 \sim R$, and to irradiation at exact resonance so that $\omega_b = 0$, the operator $\mathcal{Q}$ becomes

$$\mathcal{Q} = \begin{bmatrix} -R & 0 & 0 \\ 0 & -R & \omega_1 \\ 0 & -\omega_1 & -R \end{bmatrix}. \tag{A1.8}$$

A solution of the differential matrix equation (A1.6) with constant coefficients is (28)

$$\mathbf{m}_\rho(t) = \exp\{\mathcal{Q}t\} \cdot \mathbf{m}_\rho(0) + \exp\{\mathcal{Q}t\} \int_0^t \exp\{-\mathcal{Q}s\} R\mathbf{m}_{0\rho}\, ds \tag{A1.9}$$

The exponential $\exp\{\mathcal{Q}t\}$ is not a square matrix of the exponentials of each element of $\mathcal{Q}t$, but it can be evaluated (29) as a series in $\mathcal{Q}t$, using the eigenvalues of the matrix $\mathcal{Q}t$. The eigenvalues of $\mathcal{Q}$ are $-R$ and $-R \pm i\omega_1$ and using the procedures outlined in reference 29 we obtain

$$\exp\{\mathcal{Q}t\} = \exp(-Rt) \begin{bmatrix} 1 & 0 & 0 \\ 0 & \cos\omega_1 t & \sin\omega_1 t \\ 0 & -\sin\omega_1 t & \cos\omega_1 t \end{bmatrix}. \tag{A1.10}$$

Now if B_1 is large, so that $\omega_1 \gg R$ and we also concentrate on the short time behaviour during the pulse, then $\exp(-Rt) \sim 1$. Moreover the matrix of equation (A1.10) can be regarded as a rotation operator, $P_x(-\theta)$, about the x axis through a negative angle $\theta = \omega_1 t$.

$$\exp\{\mathcal{Q}t\} \sim P_x(-\theta).$$

The magnitude of the integral in equation (A1.9) will be governed by its coefficient which contains R/ω_1. As a result this second term can be neglected for intense pulses and long relaxation times. Under these circumstances therefore

$$m_\rho(t) = P_x(-\theta)m_\rho(0) \tag{A1.11}$$

i.e. during the pulse the rotating frame magnetization rotates around the x axis in a negative sense and at a frequency $\omega_1 = \gamma B_1$,

If a magnetization component is chosen which, due to the gradient, does not experience exact resonant irradiation, then ω_b is finite and

$$\mathcal{Q} = \begin{bmatrix} -R & \omega_b & 0 \\ -\omega_b & -R & \omega_1 \\ 0 & -\omega_1 & -R \end{bmatrix}.$$

The eigenvalues of $\mathcal{Q}$ are now $-R$ and $-R \pm i\omega_e$ where $\omega_e^2 = \omega_1^2 + \omega_b^2$, the precession frequency in the effective field $B_{\text{eff}} = \sqrt{B_1^2 + b^2}$. Following an analysis very similar to that used to obtain equation (A1.11) one obtains a new rotation operator $P(\theta_e, \psi)$ analogous to $P_x(\theta)$ of equation (A1.11) and given by

$$P(\theta_e, \psi) = \begin{bmatrix} (\sin^2\psi + \cos^2\psi\cos\theta_e) & \cos\psi\sin\theta_e & \sin\psi\cos\psi(1-\cos\theta_e) \\ -\cos\psi\sin\theta_e & \cos\theta_e & \sin\psi\sin\theta_e \\ -\sin\psi\cos\psi(1-\cos\theta_e) & -\sin\psi\sin\theta_e & (\cos^2\psi + \sin^2\psi\cos\theta_e) \end{bmatrix}.$$

where $\theta_e = \omega_e t$ and $\psi = \tan^{-1}(B_1/b)$, showing that the precession is now in a cone around the effective field B_{eff}.

References

1. A. Abragram, *Principles of Nuclear Magnetism*, Oxford University Press, London (1961).
2. P. Mansfield and P. G. Morris, *NMR Imaging in Biomedicine, Advances in Magnetic Resonance*, Supplement 2, (ed. J. S. Waugh) Academic Press, N.Y. (1982).
3. I. L. Pykett, *Sci. Am.* **246** (1982) 78–88.
4. E. O. Brigham, *The Fast Fourier Transform*, Prentice-Hall, Englewood Cliffs, N.J. (1974).
5. F. Bloch, *Phys. Rev.* **70** (1946) 460–474.
6. Reference 1, p. 45.
7. E. T. Jaynes, *Phys. Rev.* **98** (1955) 1099–1105.
8. Reference 2, p. 41.
9. R. S. Parker, I. Zupancic and J. Pirs, *J. Phys. E: Sci. Instrum.* **6** (1973) 899–900.
10. V. Bangert and P. Mansfield, *J. Phys. E: Sci. Instrum.* **15** (1982) 235–239.

11. P. Mansfield, A. A. Maudsley, P. G. Morris and I. L. Pykett, *J. Mag. Res.* **33** (1979) 261–279.
12. P. R. Locher, *Phil. Trans. R. Soc. Lond.* **B289** (1980) 537–542.
13. R. A. Brooks and G. Di Chiro, *Phys. Med. Biol.* **21** (1976) 689–732.
14. R. Gordon and G. T. Herman, *Int. Rev. Cytol.* **38** (1974) 111–151.
15. K. Kouris, N. M. Spyrou and D. F. Jackson, *Imaging with Ionizing Radiations* (*Progress in Medical and Environmental Physics*, Vol. 1) Blackie, Glasgow (1982).
16. R. B. Marr, C-N. Chen and P. C. Lauterbur, *Mathematical Aspects of Computerized Tomography*, (eds. G. T. Herman and F. Natterer) Springer-Verlag (1981) 225–240.
17. C-N. Chen, R. B. Marr and P. C. Lauterbur, *J. Mag. Res. in Medicine*, **1** (1983), in press.
18. A. Kumar, D. Welti and R. R. Ernst, *J. Mag. Res.* **18** (1975) 69–83.
19. P. Mansfield, *J. Phys. C.* **10** (1977) L55–L56.
20. P. Mansfield and I. L. Pykett, *J. Mag. Res.* **29** (1978) 355–373.
21. D. G. Taylor and C. G. Bore, *CT : J. Comput. Tomogr.* **5** (1981) 122–134.
22. H. A. Resing, K. R. Foster and A. N. Garroway, *Science*, **198** (1977) 1181–1182.
23. J. R. Zimmerman and W. E. Brittin, *J. Chem. Phys.* **61** (1957) 1328–1333.
24. W. A. Edelstein, J. M. S. Hutchinson, G. Johnson and T. Redpath, *Phys. Med. Biol.* **25** (1980) 751–756.
25. J. M. S. Hutchinson, W. A. Edelstein and G. Johnson, *J. Phys. E :Sci. Instrum.* **13** (1980) 947–955.
26. Reference 1, p. 58.
27. Reference 1, p. 103.
28. R. Bronson, *Matrix Methods*, Academic Press, New York (1969) 153.
29. Reference 28, p. 119.

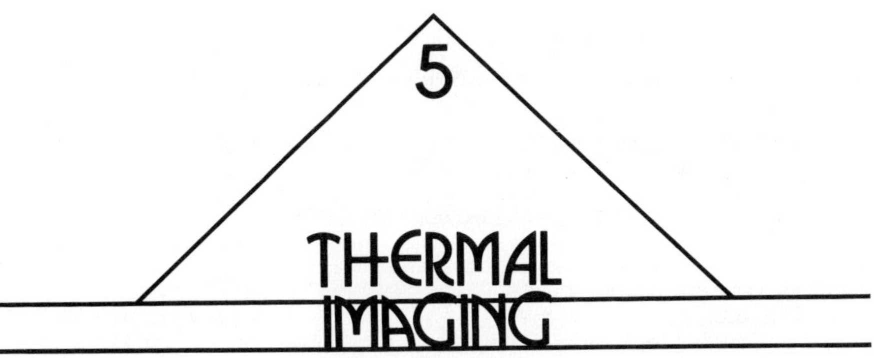

Colin H. Jones

5.1 Introduction

All objects are emitting and absorbing infrared radiation continuously. The radiant energy is emitted over a broad wavelength band with maximum emission occurring at a wavelength dependent upon the surface temperature of the object: hot bodies and fires emit at wavelengths of a few microns whereas a human body with a surface temperature of 30°C will emit most copiously at wavelengths around ten microns. Although radiant emission takes place over a wide range of wavelengths, the detectors most frequently employed in infrared thermography are sensitive to radiation of between 3 and 15 microns.

The application of electromagnetic theory to explain the variations in the emissive properties of solids is complex, and thermodynamics is usually used to predict and explain the radiative behaviour of such bodies. A full treatment of the thermodynamics of radiation is beyond the scope of this book and the reader is referred to other texts (1, 2). However, the principal concepts and results relevant to thermographic imaging are summarized below.

5.2 Thermal radiation

5.2.1 *Black body radiation*

Fundamental to an understanding of the thermal emission of solids is the concept of a black body. By definition a thermal black body is a substance which absorbs all radiation which strikes the body at any wavelength. If such a body were placed in a uniform-temperature enclosure it would come to equilibrium at the temperature of the enclosure and would therefore emit just as much radiation as it absorbs. In practice a black body may be closely approximated by a small hole in a well insulated furnace, or heated sphere. Any radiation which enters the aperture of such a sphere undergoes many low reflections on the inside surface so that all the original energy is absorbed before any radiation can be reflected back out through the aperture. Other

Table 5.1 Total emissivity values at normal incidence. Values marked * have been interpolated after Steketee (4) and are for 3–5 μm radiation.

Material	*Temperature* (°C)	*Emissivity*
METALS AND THEIR OXIDES		
Aluminum: polished sheet	100	0.05
anodized sheet, chromic acid process	100	0.55
vacuum deposited	20	0.04
Brass: highly polished	100	0.03
oxidized	100	0.61
Copper: polished	100	0.05
heavily oxidized	20	0.78
Gold: highly polished	100	0.02
Iron: cast, polished	40	0.21
cast, oxidized	100	0.64
sheet, heavily rusted	20	0.69
Magnesium: polished	20	0.07
Nickel: electroplated, polished	20	0.05
oxidized	200	0.37
Silver: polished	100	0.03
Stainless steel: type 18-8, buffed	20	0.16
type 18-8, oxidized at 800°C	60	0.85
Steel: polished	100	0.07
oxidized	200	0.79
Tin: commercial tin-plated sheet iron	100	0.07
OTHER MATERIALS		
Brick: red common	20	0.93
Carbon: candle soot	20	0.95
graphite, filed surface	20	0.98
Concrete	20	0.92
Glass: polished plate	20	0.94
Lacquer: white	100	0.92
matte black	100	0.97
Oil, lubricating (thin film on nickel base):		
nickel base alone	20	0.05
film thickness of 0.001, 0.002, 0.005 in.	20	0.27, 0.46, 0.72
thick coating	20	0.82
Paint, oil: average of 16 colours	100	0.94
Paper: white bond	20	0.93
Plaster: rough coat	20	0.91
Sand	20	0.90
Skin human: with	32	0.98
lipstick	32	0.95–0.98*
Talcum venetum	32	0.91–0.97*
Zinc oil	32	0.94*
Soil: dry	20	0.92
saturated with water	20	0.95
Water: distilled	20	0.96
ice, smooth	−10	0.96
frost crystals	−10	0.98
snow	−10	0.85
Wood: planed oak	20	0.90

bodies absorb only a fraction of the radiation falling on them and this fraction depends on the physical properties of the body and varies with the radiation wavelength. Indeed it is only in conditions when temperature equilibrium does not appertain that bodies appear different. The effective temperature of the sun is about 6000°C so in daylight at room temperature we see objects in different colours according to which wavelengths they absorb. These bodies emit radiation at room temperature and so do not replace adequately the radiation they have absorbed.

5.2.2 *Kirchhoff's law*

Kirchhoff demonstrated in 1860 that a good absorber is also a good radiator. If a body at temperature T absorbs a fraction $\varepsilon(T, \lambda)$ of monochromatic radiation of wavelength λ falling on it, then if $R_{b\lambda}(T, \lambda)d\lambda$ is the power per unit area which a perfectly black body would emit with wavelengths between λ and $\lambda + d\lambda$, the body in question will emit an amount $\varepsilon(T, \lambda)R_{b\lambda}(T, \lambda)d\lambda$. The quantity $\varepsilon(T, \lambda)$ is known as the emissivity and is a function of temperature and wavelength. For solids it varies only a little over a wide range of temperature and the variation with λ is not generally rapid. Typical values for a wide range of substances are given in table 5.1. Although radiation is emitted from all parts of a solid body, most materials are so opaque to infrared radiation that a negligible portion of the radiant power leaving the body originates more than a fraction of a millimetre below its surface.

5.2.3 *Stefan-Boltzmann law*

In 1884 Boltzmann deduced a law relating the total power emitted by a black body to its absolute temperature. If $R_b(T)$ is the total power per unit area radiated into a hemisphere by a perfect radiator then

$$R_b(T) = \sigma T^4$$

where σ is Stefan's constant and equal to $5.6686 \times 10^{-12}\,\mathrm{W\,cm^{-2}\,K^{-4}}$; R_b is called the radiant emittance of a black body and is measured in $\mathrm{Wm^{-2}}$.

For solids which are not perfectly black bodies,

$$R(T) = \varepsilon(T)\sigma T^4$$

Although $\varepsilon(T)$ is a slowly varying function of T, in most situations it can be considered as a constant. In practice bodies are not only radiating thermal energy but are being irradiated by surrounding sources. A black body whose area is A, at a uniform temperature T_1, and whose surroundings (also assumed to be thermally black) are at a uniform temperature T_2 loses heat at a rate

$$\mathrm{W} = A\sigma(T_1^4 - T_2^4)$$

If T_1 differs from T_2 by a small quantity ΔT then

$$\mathrm{W} = 4A\sigma T^3 \Delta T$$

so that for an area of $1\ \mathrm{m}^2$ which has a temperature 10°C higher than that of its surroundings the heat loss is about 60 W. If the body is not black the rate of heat loss is given by

$$\mathrm{W} = A\sigma\{\varepsilon(T_1)T_1^4 - \varepsilon(T_2)T_2^4\}$$

where $\varepsilon(T)$ is the average emissivity at temperature T.

It is the detection and quantification of such radiant heat loss that forms the basis of thermal imaging systems.

5.2.4 *Lambert's cosine law*

This states that the radiation in a small solid angle from a plane surface (or small surface element) is proportional to the solid angle and varies as the cosine of the angle between the direction in question and the normal to the surface. If the goniometric distribution of radiation emitted obeys this law, the radiation source is said to be Lambertian or perfectly diffuse. (As a consequence of this law a glowing sphere may appear to be a flat disk of uniform brightness.) Black body radiators are Lambertian but many surfaces are not and exhibit emissivities that are dependent upon direction. This has important implications in thermal imaging and will be considered later.

5.2.5 *The Planck radiation formula*

The spectral distribution of emitted radiation depends on the temperature of the body and can be represented accurately by a formula first derived by Planck. This basic law gives the energy density of the radiation in a black body enclosure and indicates that the spectral radiant emittance of a black body ($R_{b\lambda}$) into a hemisphere in the wavelength range from λ to $\lambda + d\lambda$ is:

$$R_{b\lambda} = \frac{2\pi c^2 h}{\lambda^5} \frac{1}{e^{ch/\lambda kT} - 1} d\lambda$$

where c is the speed of light
h is Planck's constant (6.6252×10^{-34} Js)
k is Boltzmann's constant (1.3805×10^{-23} J/deg K)
T is temperature in K

This equation is often written in the form

$$R_{b\lambda} = \frac{C_1 d\lambda}{\lambda^5 (e^{c_2/\lambda T} - 1)}$$

where
$$C_1 = 2\pi c^2 h (= 3.74 \times 10^{-12}\,\mathrm{W\,cm^2})$$
$$C_2 = \frac{ch}{k} (= 1.438\,\mathrm{cm\,K})$$

The spectral radiant emittance for black bodies at various temperatures is shown in figure 5.1. The area under each curve corresponds to the Stefan–Boltzmann relationship. The emittance at any particular wavelength always increases as the temperature increases. The wavelength λ_m at which the maximum radiation emission occurs decreases as the temperature increases and is expressed by the Wien displacement formula:

$$\lambda_m T = 2898\,\mu\mathrm{K}$$

The spectral radiant emittance of a black body can be written as

$$\frac{R_{b\lambda}}{T^5} = \frac{C_1}{(\lambda T)^5 (e^{c_2/\lambda T} - 1)}$$

or as

$$\frac{R_{b\lambda}}{R_m} = \frac{(\lambda_m T)^5}{(\lambda T)^5} \cdot \frac{e^{c_2/\lambda T} - 1}{e^{c_2/\lambda T} - 1}$$

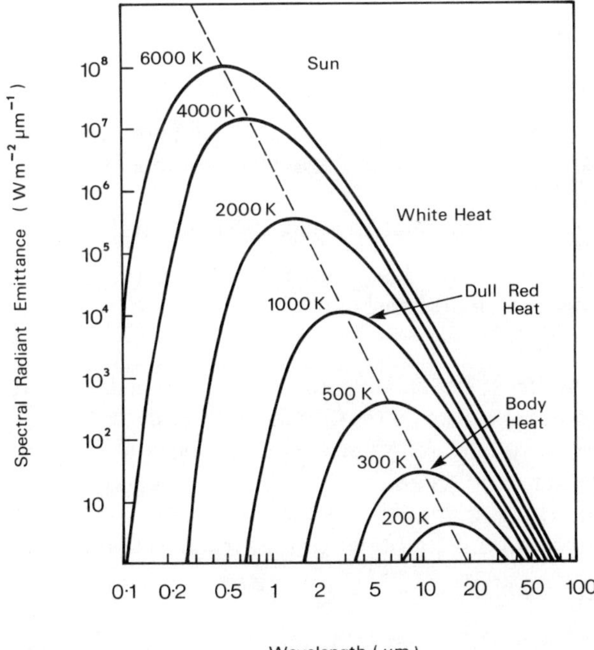

Figure 5.1 Variation of spectral radiant emission with wavelength for thermal sources at different temperatures.

where R_m is the maximum value for the spectral radiant emittance which occurs at the value of $\lambda_m T$. It follows that the ratio of the emittance at the maximum of the curve at two different temperatures equals the ratio of the temperatures to the fifth power. For example, the maximum emittance at 400°C (typical of the surface temperature found in many industrial applications of thermography) is about 40 times (i.e. 2.2^5) that at 30°C (typical of temperatures in medical work).

For some purposes it is more useful to know the number of quanta emitted by a body rather than the radiated power. It can be shown that the total number of quanta $N_b(T)$ emitted per unit area per second by a black body at a temperature T is

$$N_b(T) = 1.5213 \times 10^{15}\, T^3 \text{ quanta/m}^2$$

and may be approximated by

$$N_b(T) = \frac{\sigma T^4}{2.75\, kT}$$

which may be interpreted as showing that the average size of quanta emitted is $2.75\, kT$ (2, 5).

5.3 Principles of temperature measurement

There are three principal types of radiation source:

(1) black body sources for which $\varepsilon_\lambda = 1$ for all temperatures and wavelengths,
(2) grey body sources for which ε_λ may be considered to be a constant less than 1, and
(3) selective radiators for which ε_λ changes significantly with wavelength λ.

Radiant energy emitted by a body may be absorbed, transmitted or reflected by other objects and all of these factors may be considered to be wavelength dependent.

(a) The spectral absorbance α_λ is the ratio of the spectral radiant power absorbed by an object to that incident upon it.
(b) The spectral radiant reflectance ρ_λ is the ratio of the spectral radiant power reflected by an object to that incident upon it.
(c) The spectral transmittance τ_λ is the ratio of the spectral radiant power transmitted through an object to that incident upon it.

It follows that

$$\alpha_\lambda + \rho_\lambda + \tau_\lambda = 1.$$

According to Kirchhoff's law, for any material ε_λ and α_λ are equal at any specified temperature and wavelength, so

$$\varepsilon_\lambda + \rho_\lambda + \tau_\lambda = 1.$$

For opaque materials $\tau_\lambda = 0$ so that

$$\varepsilon_\lambda + \rho_\lambda = 1.$$

Remote temperature measurement by an infrared camera must allow for the fact that the infrared radiation received by the detector might be composed of three separate components (6). These are:

(1) radiation emitted by the object as a result of the surface temperature of the object (T_0) and its emissivity (ε),
(2) radiation reflected by the object as a result of the ambient temperature of the surroundings (T_a); this will be proportional to the reflectance (ρ) of the object, and
(3) radiation transmitted through the object as a result of either the background temperature, or temperature (T_d) at a depth within the body of the object.

These three radiation components are interdependent and can be written as:

$\varepsilon f(T_0)$ = emitted radiation which is a function of the object's surface temperature

$\rho f(T_a)$ = reflected radiation which is a function of the ambient temperature

$\tau f(T_d)$ = transmitted radiation which may be a function either of the background temperature, or transmitted radiation from deep within the object.

In practice, most objects are non-transparent and for the wavelengths with which we shall be dealing the transmitted contribution can be neglected. The signal S resulting from a small quantity of the total radiation emitted by an object impinging upon the detector in a thermographic system can be written

$$S = \varepsilon f(T_0) + \rho f(T_a)$$

since $\rho = 1 - \varepsilon$ then

$$S = \varepsilon f(T_0) + (1 - \varepsilon) f(T_a)$$

The camera-signal *v*. black body temperature characteristic curve $S_b = f(T_b)$ depends upon both the type of detector and the system optics, but is often exponential in form. Fortunately many surfaces radiate with a spectral distribution similar to a black body, though with smaller absolute values of intensity (so called "grey" radiators) and if ε is the surface emissivity then

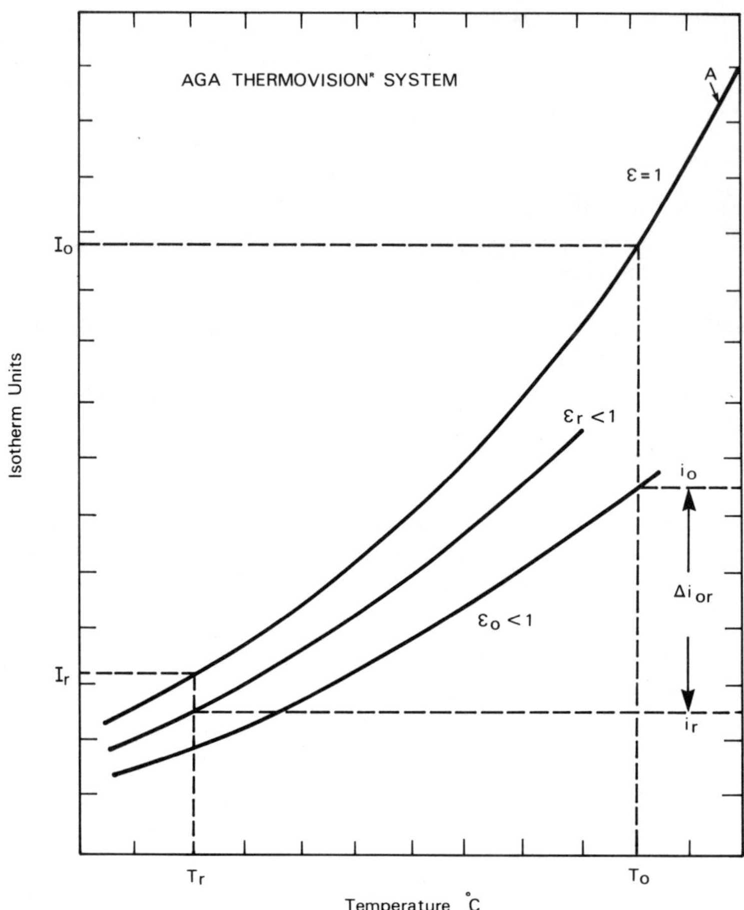

Figure 5.2 Calibration curves for an Aga Thermovision infrared camera system.

the corresponding characteristic response curve for the detector will be $S_0 = \varepsilon f(T_b)$. Manufacturers usually provide calibration curves for various aperture settings, either in signal voltage or isotherm units. In figure 5.2 curve *A* represents a typical calibration chart for an Aga Thermovision camera with an InSb detector, and shows the relationship between the true object and reference temperature T_0 and T_r respectively, and the absolute isotherm levels I_0 and I_r for a black body ($\varepsilon = 1$). On the video display of the system, the thermal images of the object and reference sources are related to the isotherm levels i_0 and i_r by the ε_0 and ε_r curves respectively: i_0 and i_r correspond to isotherm levels obtained from the picture display and depend upon the selected sensitivity and black level controls.

The camera response signal may be written in the form

$$S_0 = \varepsilon_0 I_0 + (1 - \varepsilon_0) I_a \text{ for the object temperature and}$$
$$S_r = \varepsilon_r I_r + (1 - \varepsilon_r) I_a \text{ for the reference temperature}$$

It is also evident that the differences between isotherm levels $(i_0 - i_r)$ can be equated with differences between signal levels $(S_0 - S_r)$ so:
if

$$\Delta i_{0r} = i_0 - i_r = S_0 - S_r$$

then

$$\Delta i_{0r} = \varepsilon_0 I_0 - \varepsilon_r I_r + (I - \varepsilon_0) I_a - (1 - \varepsilon_r) I_a$$

and

$$I_0 = \frac{\Delta i_{0r}}{\varepsilon_0} + \frac{\varepsilon_r}{\varepsilon_0} I_r + \left(1 - \frac{\varepsilon_r}{\varepsilon_0}\right) I_a$$

This formula may be used to determine I_0 and then T_0, the true temperature of an object, providing the object's emissivity ε_0, reference temperature T_r and reference emissivity ε_r are all known.

The following generalizations concerning the emissivity of surfaces are helpful (5):

1. The emissivities of non-metallic substances are typically greater than 0.8 at $T = 350$ K; the emissivity decreases with temperature, and values are typically between 0.3 and 0.8 at higher temperatures.
2. The emissivities of most metals are very low and are approximately proportional to the absolute temperature; the proportionality constant varies as the square root of the electrical resistance at a fiducial base temperature.

The emissivity of a surface is affected principally by surface structure, and consequently surface contaminants or oxidation can alter the emissive properties of a material. Allowance must be made for this possibility whenever infrared techniques are used to measure temperature magnitudes. If the surface is at an absolute temperature T_0 and has an emissivity ε_0 the radiant power emitted from the surface will be proportional to $\varepsilon_0 T_0^4$. Calibration of a radiometer may be achieved by measuring the output when it views a black body of known temperature T_b. The corresponding power will be proportional to T_b^4 so:

$$T_b = \varepsilon_0^{1/4} T_0$$

Thus a 1% change in emissivity causes an apparent reduction in temperature of about 0.75°C. In the case of photon detectors the effect of changes in emissivity on the accuracy of temperature measurement is complex and depends upon the spectral sensitivity of the detector and the temperature of the object (7). For example, the spectral emissivity of human skin is 0.98 ± 0.01 (8) for wavelengths 1–20 μm but this value can change when the skin is covered with cosmetics with a resultant decrease in radiant emission which may be interpreted erroneously as a lower temperature. Steketee (4) has shown that the apparent temperature decrease when measured with InSb detectors sensitive to 2–5.5 μm radiation is about 0.3 K for talc venetum, whereas for Hg Cd Te detectors the temperature difference is 1.3 K due to the greater spectral sensitivity at 8.5 μm and the variation ε with wavelength.

Consideration must also be given to the effects of emission angle on emissivity values. Although most of the surfaces encountered in thermal imaging can be considered to be "grey" radiators, data obtained by viewing a surface at an angle $\phi > 60°$ to normal incidence (ϕ_0) must be corrected to allow for emissivity variations with direction. In the case of electrical insulators electromagnetic theory shows that

$$\varepsilon_{0\lambda} = 1 - \left(\frac{n_\lambda - 1}{n_\lambda + 1}\right)^2$$

where $\varepsilon_{0\lambda}$ is the emissivity ($\phi = 0$) for wavelength λ.

$n_\lambda = \dfrac{\sin\phi}{\sin x}$ is the coefficient of refraction.

ϕ is the angle of incidence for a ray passing through air into a medium optically denser than air and x is the angle of refraction in this medium.

It can be shown that for long wavelength radiation for an angle of incidence or emission ϕ, the emissivity of an insulator is given by Fresnel's formula:

$$\varepsilon_{\phi\lambda} = 1 - \frac{1}{2}\frac{\sin^2(\phi - x)}{\sin^2(\phi + x)}\left[1 + \frac{\cos^2(\phi + x)}{\cos^2(\phi - x)}\right]$$

The form of this function is seen in figure 5.3 which shows the goniometric variation of emissivity for glass ($n_\lambda = 1.5$), paper and copper oxide. For values of ϕ up to about 60° the curves approach the shape of a circle, but at greater angles the emissivity ε_ϕ decreases considerably and at $\phi = 90°$ it becomes zero. Thus Lambert's law holds only approximately. The emissive properties of electrical conductors can also be found theoretically from electromagnetic theory (5), but the problem is somewhat more complex. For our purposes it is sufficient to note that the radiation has a minimum for $\phi = 0°$ and increases with an increase of ϕ. For metals, the total emissivity becomes 1.33 to 1.05 times the normal emissivity, whereas for electrical insulators the corresponding ratio is between 1.05 and 0.95.

Watmough *et al.* (9, 10) and Clark (11) have considered the implications of viewing obliquely on clinical thermography. Rearrangement of the equation

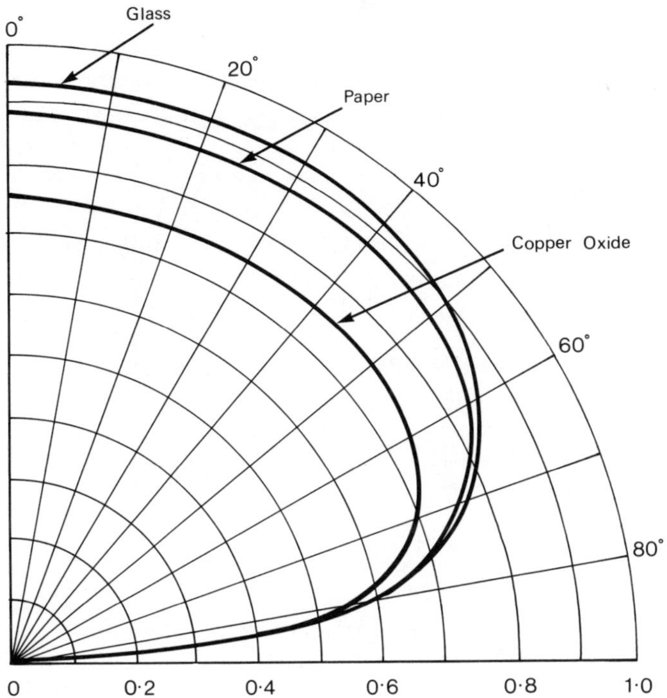

Figure 5.3 Goniometric distribution of infrared radiation emitted from copper oxide, paper and glass surfaces (5).

above gives

$$\varepsilon_{\phi\lambda} = 1 - \frac{1}{2}\left\{\frac{\beta - \cos\phi}{\beta + \cos\phi}\right\}^2 \left\{1 + \left[\frac{\beta\cos\phi - \sin^2\phi}{\beta\cos\phi + \sin^2\phi}\right]^2\right\}$$

where

$$\beta = (n_\lambda^2 - \sin^2\phi)^{1/2}$$

Figure 5.4 shows this function plotted for a smooth surface with $\varepsilon_{0\lambda} = 0.98$ at normal incidence which corresponds to human skin at wavelengths typical of those used thermographically. Watmough *et al.* (9) have shown that over a wide range of emissivities the apparent emissivity of a curved surface varies very little for angles of view between the normal and $\pi/4$ but at greater angles the emissivity changes significantly. In practice allowance must also be made for the reflectance of radiation from the surroundings by the surface being imaged. This becomes more significant as the emissivity of the object diminishes. Consequently the measured change in temperature with viewing angle due

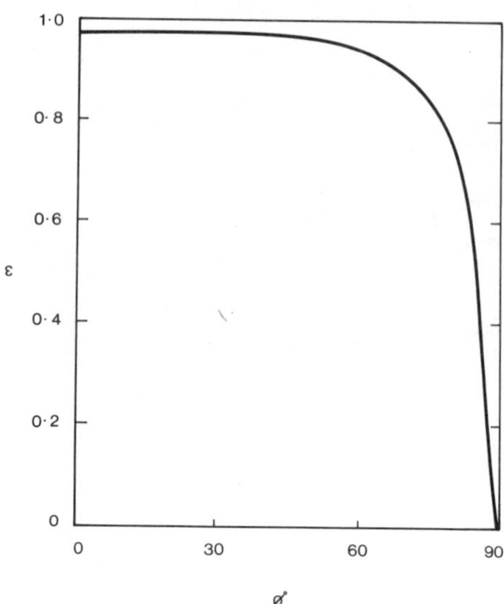

Figure 5.4 Variation of emissivity ε with angle of view ϕ predicted by theory for a smooth surface with $\varepsilon = 0.98$ at normal incidence (11).

to a reduction in emissivity is less than theoretical predictions and depends upon the temperature of the environment (11).

Thermography frequently requires the use of a plane mirror for viewing the scene. Conventional glass-fronted mirrors are unsuitable for infrared imaging and it is customary to use highly reflective front silvered mirrors. For a system with a cooled indium antimonide detector, highly polished aluminium or aluminized Mylar can be used as an inexpensive, robust reflector. The reflectivity of a mirror will affect the apparent temperature of the objective: allowance must also be made for the viewing angle. For an object temperature of 32°C, at viewing angles up to 45°, copper and silver show an apparent temperature reduction of 0.2°C whereas aluminium and aluminized Mylar show an apparent temperature reduction of 0.5°C.

In medical thermography it is usually assumed that skin emissivity approaches unity and is constant throughout the relevant temperature range. In industrial and environmental thermography careful attention must be paid to the emissive properties of the surface being investigated and appropriate ε values must be used if quantitative measurements of temperature are to be determined.

This limitation is not as serious as it might appear for two reasons: firstly, absolute temperature measurements are not always necessary—temperature differences are sufficient for resolving many problems—and secondly, the emissivity of most surfaces changes relatively slowly with temperature.

5.4 Thermographic detectors

There are two groups of instrument that can be used to detect infrared radiation: those which use the heating effect of radiation which are known as thermal detectors, and those which make use of quantum photo-electric effects which are usually referred to as photon detectors. Radiation thermocouples, bolometers and the Golay cell are all examples of thermal detectors, whereas photon detectors include devices that depend upon the photoemissive, photoconductive or photovoltaic properties of certain materials (2, 5, 7).

5.4.1 *Thermal detectors*

Thermal detectors sense radiation by the temperature rise in the absorbing element which in turn affects some temperature-sensitive property. For example, semi-conducting materials with high thermoelectric powers and low thermal conductivity are used in Schwarz-type thermocouples which were employed in some of the earliest clinical thermographic studies to detect breast cancer (12).

An elegant approach to infrared detection devised by Czerny makes use of the variation in evaporation rate with temperature of an oil film. This principle was developed by Baird Atomic Inc., and incorporated into a thermal imager called the Evaporograph (13). The detector cell consists of an evacuated chamber in which is mounted a thin membrane coated with gold black on the outside surface. The chamber temperature is adjusted so that a thin oil film is established on the membrane surface in equilibrium with warm oil vapour in the chamber. Radiation enters the chamber through a rock-salt window and forms a thermal image on the blackened side of the membrane. This causes the oil film to evaporate at a rate proportional to the local temperature rise so the condensation rate is more rapid on the cooler parts of the image. The change in oil film thickness is seen as a variation in the visible interference pattern reflected by the membrane. After several seconds the image can be photographed and the picture colours can be calibrated according to temperature differentials.

Although now used infrequently, thermistor bolometer systems were developed for thermal scanning and have been used successfully for clinical thermography (14). A bolometer employs an absorbing element (usually made from a combination of the oxides of nickel, manganese and cobalt) whose electrical resistance is temperature-sensitive. By use of a steady bias current, the change in resistance caused by the temperature change produces a voltage signal which can be amplified electronically. Detectors of this type have time constants of the order of milliseconds and they have a lower frequency response than photon detectors. The high value of the current noise in a bolometer limits the ultimate sensitivity of the detector and they are less sensitive than photon detectors.

5.4.2 *Photon detectors*

Photo-electric cells can be made to detect radiation at wavelengths up to 1.2 μm, but such devices have been used mainly for the detection of radiation in the visible region of the spectrum. Infrared sensitive cathodes have maximum sensitivity at 0.8 μm, and the sensitivity falls to 1% of this value at 1.2 μm. Such devices are not used in thermal imaging.

Photon detectors used for imaging of 2–15 μm wavelength radiation are usually semi-conductor type devices in which the initial absorption of a photon results in the freeing of bound electrons or charge carriers. Whereas crystalline solids that are classified as insulators have forbidden energy gaps that are greater than about 2 electron volts ($1\,\text{eV} = 1.6 \times 10^{-19}$ joule), the forbidden band in semi-conductors ranges from 0.01 of an electron volt to a few electron volts. Semi-conductors may be divided further into two general categories called intrinsic and extrinsic (impurity) semi-conductors. The intrinsic form has no additional energy levels in between the valence and conduction bands. Electron excitations can take place only when the absorbed photon energy is sufficient to traverse the forbidden gap. Electrons in the valence band are excited up to the conduction band, leaving holes in the valence band. One of the most frequently used intrinsic detectors is indium antimonide. The width of the forbidden band in InSb is 0.18 eV at room temperature corresponding to a cutoff wavelength of about 7 μm; on cooling to 77 K the band width increases to 0.23 eV giving rise to a cutoff wavelength of 5.5 μm. An extrinsic semi-conductor is not a pure, perfect crystal but contains imperfections such as impurities, lattice defects or dislocations, with additional energy levels which occur within the forbidden gap. In general, the electrons located in these levels are trapped in position but they may be excited from the impurity level to the conduction band, or from the valence band to the impurity level. Thermal excitation will remove some of the electrons from the filled impurity levels and fill some of the empty impurity levels from the valence band. The crystal will thus have either an excess of electrons in the conduction band (*n*-type semi-conductor) or an excess of holes in the valence band (*p*-type semi-conductor). By doping a semi-conductor with an impurity it is possible to manipulate its response to infrared radiation. For example, Ge is not used for thermal imaging because its wide band gap of 0.67 eV corresponds to a cutoff wavelength of about 1.85 μm. By adding impurities soluble in Ge it is possible to produce "extrinsic" semi-conductors which have shorter cut-off wavelengths. Gold-doped germanium for instance when cooled to 77 K responds to radiation up to about 8 μm.

The search for detectors with energy gaps small enough to permit operation beyond 10 μm led to the formation of mixed crystals of mercury telluride and cadmium telluride. Pure HgTe has a gap width of 0.03 eV whereas pure CdTe has a gap width of 1.5 eV. Lawson *et al.* (15) developed a technique for producing mixed crystals of varying composition from pure CdTe at one extreme to pure HgTe at the other. By varying the ratio of the components, intrinsic semi-conductors may be produced with energy gaps corresponding to any desired value up to 15 μm. This range includes the important 8–13 μm atmospheric window and also the peak spectral emissive power around 10 μm for clinical thermography. Detectors of this type are described as $Hg_xCd_{1-x}Te$, where x refers to the percentage of Hg in the crystal. HgCdTe detectors are frequently referred to as CMT detectors. Mixed crystal detectors have also been manufactured from lead telluride and tin telluride.

In practice there are two principal photo-effects that are employed in semi-conductor type devices: these are the photo-conductivity and the photovoltaic effect. When used in the photoconductive mode the high electrical resistance of the sensor is decreased on illumination with infrared radiation. In practice the photo-conductive detector forms part of an electrical circuit in series with a load resistor and a battery. Infrared radiation which is modulated or "chopped" (for example, by a rotating disc or fan wheel) is absorbed by the photo-detector causing its resistance to decrease. The resultant potential changes across the load resistor are then impressed upon an amplifier circuit.

The photovoltaic effect is an internal photo-effect in which the action of photons within a semi-conductor produces a voltage which can be detected directly without need for bias supply or load resistor. In the absence of radiation falling upon the junction, the Fermi levels in the *p* and *n* regions are aligned. A photon of sufficient energy to cause intrinsic excitation, when absorbed at the junction produces a free hole-electron pair. Under the influence of the applied electric field the electron moves into the *n*-type material and the hole into the *p*-type material. As long as radiation falls upon the junction, electron-hole pairs will be formed and separated by the electric field at the junction. If the semi-conductor ends are short-circuited by an external conductor, a current will flow in the circuit; if the ends are open-circuited by a high impedance voltage measuring device, a voltage will exist as long as radiation falls upon the sample.

The photoelectromagnetic effect (PEM) can also be employed to detect

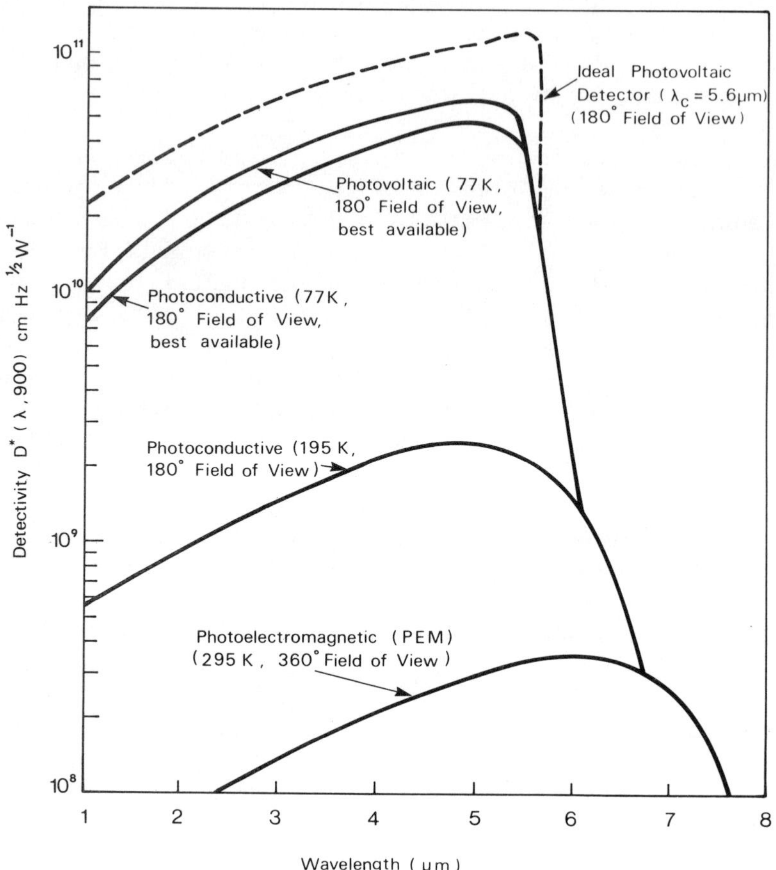

Figure 5.5 Variation of detectivity D^* with wavelength for InSb detectors used in different modes (3).

infrared radiation, but this is not often used for thermal imaging. This is an internal photo-effect which is dependent upon the diffusion of electrons and holes caused by radiation falling upon a semi-conductor in an applied magnetic field.

The spectral response and characteristics of photon detectors depend upon the mode of operation. Figure 5.5 shows the spectral detectivity of InSb detectors used in various modes (3). Most thermographic imaging systems have employed cooled InSb or HgCdTe detectors in photo-conductive or photovoltaic modes. Because the output of a photon detector does not depend upon a rise in its temperature, its response time depends on the photo-electronic properties of the detector and is very fast (typically a few microseconds) making it particularly suitable for fast scanning systems. For most applications it is necessary to cool the detector and employ complex, high quality optics and electronics to form thermal images. These subjects are considered comprehensively by several authors (3, 5, 7).

5.4.3 *Pyroelectric infrared systems*

Alternative less expensive detectors have also been developed including those based on the pyroelectric effect which is exhibited by certain ferro-magnetic crystals such as barium titanate and triglycine sulphate (TGS). Electrically the sensor behaves as a capacitor on which a charge appears when it is exposed to a change in radiance. The magnitude of this polarization depends upon the rate of temperature change in the detector so the material does not respond to a steady flux of radiation. The effect occurs at room temperature, so cryogenics are not required. A further advantage is that the detector responds to changes in flux over a large range of wavelengths.

Relatively inexpensive pyroelectric scanning systems have been developed commercially (16, 17). There are three ways in which detectors have been used:

(1) a single detector used in conjunction with a two-dimensional mechanical scan of the scene;
(2) a linear array of detectors used with a one-dimensional scan;
(3) a two-dimensional detector matrix.

Burgess has described the technical aspects of these systems (18). In the third case, mechanical scanning may be avoided by using a pyroelectric vidicon camera tube in place of the detector array (19). By means of a germanium infrared transmitting lens, the image of the thermal scene is focused on to a thin disc of pyroelectric material (20 mm diameter, 30 μm thick) causing temperature differences which induce a charge pattern. This pattern is scanned in a TV raster by the electron beam of the vidicon tube. Since a signal change is produced only when the target temperature changes, modulation of the incident radiation is necessary. This may be accomplished either by chopping or panning of the thermal scene.

The pyroelectric vidicon forms the basis of a simple thermal imaging camera which can be coupled directly to a standard picture monitor and video recorder. The faceplate and focusing lens are usually made of germanium to allow transmission of infrared radiation and typical systems are sensitive to 8–12 μm radiation. Figure 5.6 illustrates the basic features of a typical system. Although pyroelectric vidicons are capable of resolving temperature differences of 0.1K at low spatial frequencies, at spatial frequencies similar to those encountered in clinical problems the thermal

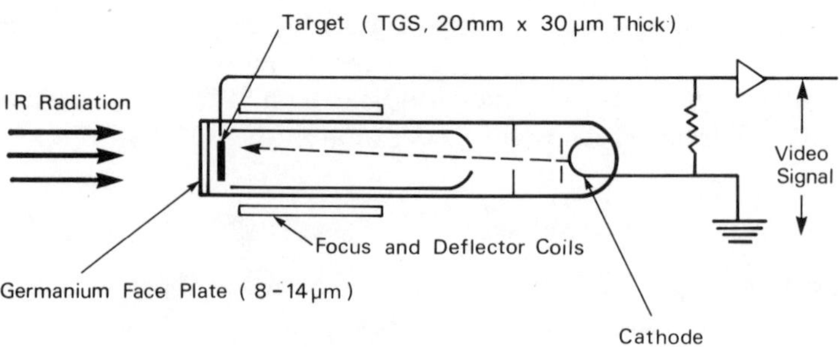

Figure 5.6 Basic components of a pyroelectric vidicon infrared camera tube.

resolution is reduced to about 0.4 K. Recently, improved camera tubes have been developed. Loss of resolution in the pyroelectric vidicon is in part due to thermal diffusion in the target. A higher resolution is obtainable by dividing the target into a matrix of thermally isolated islands stuck on to a supporting plastic backing. This process, known as reticulation, has increased tube resolution from 250 to 350 picture points (19).

Pyroelectric vidicons do not require liquid nitrogen cooling and are significantly cheaper than photon type detector systems. They have been used for medical investigations in which high thermal and spatial resolution are not required (20). Vidicon systems are likely to prove particularly useful for many industrial and military applications.

5.4.4 Detector performance

A commonly used figure of merit is the noise equivalent power (NEP). This is defined as the rms value of the sinusoidally modulated radiant power falling upon a detector which will give rise to an rms signal voltage (V_s) equal to the rms noise voltage (V_n) from the detector. Noise may be due to thermal and electrical fluctuations and it is necessary to specify the frequency of the measuring system. The temperature of the black body radiation source must be specified and is usually 500 K; the reference bandwidth is commonly 1 Hz and 5 Hz and the centre frequency is normally 90 Hz, 400 Hz, 800 Hz or 900 Hz. The noise equivalent power for a 500 K source, 900 Hz chopping frequency and a 1 Hz bandwidth is written as NEP (500 K, 900, 1). The units of NEP are watts/(Hz)$^{1/2}$ but it is common practice to omit the (Hz)$^{1/2}$. Another figure of merit describing detector performance is the detectivity, D, which is simply the reciprocal of the NEP:

$$D = \frac{1}{\text{NEP}}.$$

For many photon detectors the NEP is directly proportional to the square root of the area of the detector. To overcome this difficulty a normalized detectivity D^* is to be preferred and is used universally. The specific detectivity is defined as

$$D^* = DA^{1/2}(\Delta f)^{1/2} = \frac{\frac{V_s}{V_n}[A(\Delta f)]^{1/2}}{\mathrm{W}}$$

where A is the area of the detector and Δf is the frequency band width of the measuring system, and W is the radiation power incident on the detector (rms value in watts). The reference bandwidth is always 1 Hz so D^* for a detector is written in the form D^* (500 K, 900, 1) and is measured in $\mathrm{Hz}^{1/2}W^{-1}$. Whereas thermal detectors respond uniformly to all wavelengths, photon detectors have spectral responses characteristic of the detector and NEP, D, and D^* can be expressed in terms of their response to monochromatic radiation NEP_λ, D_λ and D^*_λ respectively. Figure 5.7 and table 5.2 show values for a variety of detectors (21, 22). The spectral responsivity of a detector, which is defined as the rms signal voltage per unit rms radiant power is specified usually in terms of peak wavelength and chopping frequency of the measuring system. For InSb, at 5.3 μm and 800 Hz the responsivity is typically 3×10^4V/W.

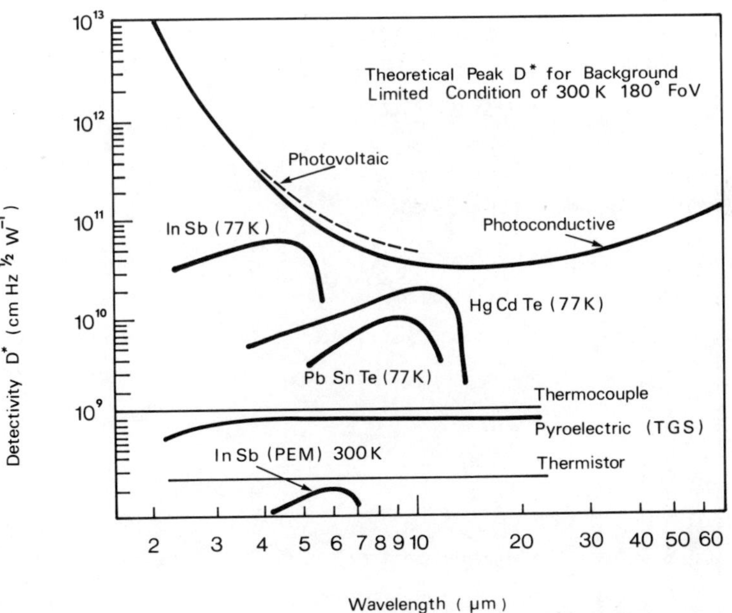

Figure 5.7 Variation of detectivity D^* with wavelength for thermal detectors.

Table 5.2 Infrared thermography systems: parameter-component descriptions.

Parameter-component	*Range of values, description*	*General information*
Spatial resolution	0.75–2 milliradians	Close-up lens can be used to improve resolution
Number of horizontal lines (n)	90–625	Large n and e necessary for vascular resolution
Elements per line (e)	100–600	10 000–360 000 elements per frame
Field of view	5° × 5°–40° × 40°	Various lens options possible
Depth of focus (d)	1–30 cm	
Range of focus (T)	10 cm to infinity	Small T useful for close-up work
Frame rate (f)	0.3 min^{-1}–60 s	Low f gives flicker picture, difficult to view and focus, requires long persistence CRT or storage facilities
Noise equivalent temp. diff.	0.07–0.3°C at 30°C	Governs temperature resolution 0.1–0.4°C
Temperature range (t)	25–45°C (medical)	2, 5, 10, 15°C ranges useful for medical thermography
	− 20–900°C (industrial)	Additional filters extend range to + 2000°C
Detector	InSb(2–5.6 μm, 77 K)	Proven reliability but narrow spectral response
	CdHgTe(2–14, 77 K)	Wide spectral response
	PbSnTe(2–14 μm, 77 K)	Wide spectral response
	Ge(Hg)(2–14 μm, < 30 K)	Used in dual systems with InSb: cooling difficult
	Multiple sensor arrays of InSb or CdHgTe	Detector matching necessary, high f, good resolution
	TGS pyroelectric Vidicon (room temperature)	Inexpensive, inferior resolution, TV compatible
	Thermistor bolometer (room temperature)	Less sensitive than photon detectors
Scanning mechanisms	Oscillating mirrors;	Low f, too slow for dynamic studies.
	Rotating prisms;	Compact, portable systems, high optical efficiency
	Rotating mirror drum with offset faces;	Flicker free, good for dynamic studies
	Pyroelectric Vidicon tube with chopper	Flicker free, real time, TV compatible, poor resolution
Scan configurations	Linear scan—single detector	Simple electronics, low f
	Parallel scan—multiple detectors in line	Each line scanned by a different detector; high channel uniformity required: low f, low channel bandwidth; complex electronics; frame store for TV compatibility.
	Serial scan—multiple detectors in line	Each line scanned by all detectors; good picture uniformity; high f; high channel bandwidth; standard electronics and TV.
	Matrix scan—multiple detectors in matrix. Several SPRITE detectors in parallel or multiple detectors in staring array.	Serial component improves uniformity; high f; complex electronics; TV compatibility.
Display systems and image processing	CRT, photography, cine-photography	Isotherms, colour, thermal amplitude analysis, area integration, digital recording, computer compatibility.

Theoretically, the detectivity of a detector may be improved to the point where the noise level is set by phenomena occurring external to the detector. When such a situation is reached further manipulation of the detector characteristics will not improve the detectivity. Apart from sources of electronic noise the principal noise source is radiation from the detector surroundings. Random fluctuations of photon flux incident on the detector from the background and target as well as photon noise in the radiation emitted by the detector to the surroundings all contribute to the resultant noise level. The nature of these random fluctuations and their effect on infrared systems are considered comprehensively by several authors (2, 3, 5). The condition of operation of a detector when limiting noise is solely due to incident photon fluctuations has been termed the BLIP (for Background-Limited Infrared Photoconductor). The specific detectivity of a detector operating under the BLIP condition (D^* BLIP) is defined as the specific detectivity of a detector of responsive quantum efficiency of unity over its spectral range of sensitivity, whose noise level results directly from the photon fluctuation of the incident flux from the background. D^* BLIP varies with the temperature of the background, the solid angle subtended by the background at the detector and the long-wavelength cutoff of the detector.

In practice the detector forms part of a measuring system comprising many complex components and detector performance is only one of many factors which affect image quality. Comparison of detector systems is aided by a measure of the noise equivalent temperature difference (NETD). This parameter is defined as the temperature difference between two black bodies that gives a signal that is equal to the total noise amplitude of the detection system. A typical thermography machine has an NETD of about 0.1 to 0.2°C (see table 5.2). Factors influencing the design and construction of thermographic systems have been described by Wolfe (21) and Lawson (23).

The angular resolution of such a system depends largely on the response time of the detector and the characteristics of the scanning optics. This quantity may be defined as the angle in milliradians subtended by an observed object which is small enough to reduce the video signal of the system to half of the maximum signal amplitude obtained for large angles. Values range from 1 milliradian for a CMT detector with a 2-second scan time to 3 milliradians for an InSb detector with an 0.04-second scan time.

5.4.5 *Liquid crystals*

Liquid crystals are a class of organic compounds which possess ordered fluid phases and combine solid-like optical properties with fluid-like flow. Some liquid crystals are colour-temperature sensitive and can be used to visualize surface temperature distributions. Most temperature studies have been carried out with cholesteric liquid crystals which are colour-temperature sensitive in the cholesteric phase (24). This phase exists within a specific temperature range and is exhibited by many esters of cholesterol (such as cholesteryl nonanoate) as a state of matter with an ordered molecular arrangement intermediate between a three-dimensional solid and a liquid. The thermochromic properties of liquid crystals arise from the ability of uniformly aligned helices existing within the cholesteric phase to selectively reflect incident radiation, the wavelength (λ) of which is related to the helical pitch (p) by $\lambda = \bar{n}p$ where $\bar{n}$ = the average refractive index. The pitch length is sensitive to temperature in the region of the first order phase transition from the less mobile smectic phase (in which molecules are ordered in layers) to the cholesteric phase. If the pitch length is of the order of the wavelength of visible light then thermochromism is observed. When viewed on a black background the scattering effects within the material give rise to iridescent colours, the dominant wavelength being influenced by very small changes in temperature. The high temperature sensitivity makes liquid crystals useful for thermal mapping and the method has been used both for medical and for industrial purposes (25, 26).

Mixtures of liquid crystals can be prepared to respond at specific temperature intervals (normally about 3°C) selected from a broad range of temperatures. Liquid crystals respond rapidly and reversibly to temperature changes. This makes them particularly suited to thermal stress investigations in which test surfaces painted with liquid crystals can be photographed for subsequent analysis. For a specific mixture each colour viewed by the observer corresponds to a particular temperature. Although very high spatial and temperature resolutions have been claimed (40 lines/mm, and 0.1°C) with a time constant of 0.1 s, these can only be achieved with spectrophotometer calibration and careful experimental conditions (24, 27). In practice the surface to be examined must be painted black before applying the liquid crystals and for clinical investigations this can be inconvenient (28).

Some of the disadvantages of painting are obviated by liquid crystal plate (or sheet) thermography (29, 30). The temperature-sensitive plate consists of a blackened thin film support into which cholesteric liquid crystals microencapsulated in gelatine (with particle sizes between 10 and 30 μm) have been incorporated. Thermal contact between the warm surface and the plate produces a colour change in the encapsulated liquid crystals; usually red corresponds to relatively low temperatures and other colours of the visible spectrum to violet for high temperatures. For some applications the plates are kept taut by a supporting frame or alternatively strips of liquid crystal sheet can be used to cover the surface to be examined (see Plate I, A). Encapsulated liquid crystals have also been incorporated into elastomeric Flexi-Therm material which facilitates vacuum contouring of non-flat surfaces

(31). The absolute temperature range of response of each plate (sheet, or strip) depends upon the liquid crystal constituents but covers a range of about 3°C; some systems contain two layers of liquid crystal to increase the range and sensitivity. Thermal diffusion within the plate causes loss of thermal and spatial resolution of the temperature pattern but qualitative comparisons of infrared and liquid crystal plate thermography show a close correspondence between the recorded images. The plate thickness varies considerably between commercial systems and ranges typically from 0.06 mm to 0.3 mm with response times of 20 to 40 seconds.

Chiral nematic thermochromic liquid crystals have also been advocated for thermographic studies (32). These optically active crystals have been synthesized so that the structure of the core of the molecule may be manipulated to extend its range of properties. Preliminary studies suggest that plates incorporating chiral nematic liquid crystals are more accurate, more stable and respond more rapidly than cholesteric encapsulated liquid crystals. Improvements in resolution have also been noted.

Whatever type of liquid crystal is used, plate thermography suffers from two major disadvantages. Firstly, it is a 'contact method' of temperature measurement, and pressure between the flexible plate and the object being investigated can influence the very temperature it is being used to measure. To some extent this difficulty can be overcome when used clinically by warming the plate to body temperature before making a measurement. Secondly, it is not always possible to obtain uniform contact between the plate and the object. This latter aspect is particularly important in clinical studies. The method has been used extensively as an inexpensive alternative to infrared thermography in breast cancer studies, investigating back pain, sports injuries and veterinary examinations (35, 36, 37, 38).

5.5 Thermographic systems

A thermographic system consists of three basic parts: (1) the scanning system which views the scene to be imaged and focuses the infrared radiation on to (2) a detector which in turn is linked to (3) a display system. The signal output from the detector is amplified and used to modulate the intensity of the electron beam of a TV type display unit. The resulting thermal image usually shows relative temperature differences in a continuous range of grey tones from black to white although some systems present the image in colour or in a digitized format. The hot areas may be displayed as white or black (inverted mode) depending on the preference of the user. A photographic record of the displayed image is usually required and facilities for obtaining these thermograms should be of high quality if image degradation is to be avoided.

Numerous scanning methods have been devised and used in conjunction with various detector configurations for scanning of the scene viewed by the detector. For single detectors scanning systems include the use of oscillating mirrors, rotating prisms and rotating multi-sided mirror drums. The build-up of the thermographic image is synchronized precisely with the scanning of the patient. The resultant thermal image is usually displayed on a television image tube. The Aga Thermovision camera is typical of a single detector instrument. In the 680 Medical version of this machine, a germanium lens is used to focus infrared radiation onto a cooled indium antimonide detector through two eight-sided prisms. A virtual image is formed by the lens on a plane within the first prism. The image is scanned vertically by rotation of the prism about its horizontal axis, which results in a horizontal virtual line-image being formed within the second scanning prism. The line image is then scanned horizontally in turn by rotation of the second prism about its vertical axis. The final image comprises 140 lines (210 elements/line) and is produced at a frame rate of 16 frames/second. The optics of the system are such that with an indium antimonide detector with a sensitive area of typically 0.3×0.3 mm used in photovoltaic mode the temperature resolution is about 0.2°C. Table 5.2 shows details of scanner parameters with a description of some of the advantages of various systems (7, 22, 23, 39).

Most commercial thermography systems employ either indium antimonide or cadmium mercury telluride detectors used in photoconductive or photovoltaic mode. Time constants are in the range 2 to 5 s for InSb detectors and 0.2 to 1 s for CMT detectors. The majority of medical and industrial work has been carried out with single-element detectors but more recent developments have resulted in the fabrication of arrays of multiple small elements (40). The output signal of InSb and CMT detectors is very small at temperatures about 20°C and is swamped by thermal noise due to random generation and recombination of electron-hole pairs in the semiconductor. However, by cooling to cryogenic temperatures it is possible to make the detector sensitive enough to resolve small temperature differences in the scene being imaged and to make the response clearly distinguishable from thermal noise. Temperatures about −196°C are usually used, but InSb detectors are sometimes used at about −80°C. Typical cooling systems consist of a precision-bore dewar into which liquid nitrogen is fed or some form of cooler is encapsulated. A Joule-Thomson cooler using air at high pressure can cool a detector to −196°C in about 30 seconds and consumes about 80 W per watt of cooling load. A Stirling-cycle cooling engine requires about 5 to 10 minutes cool-down time to −196°C and consumes about 50 W per watt of cooling load. Thermoelectric (Peltier effect) coolers may be employed in systems requiring cooling to −80°C; these have cool-down times of about 25 seconds and use only a few watts (41). The infrared detector is mounted

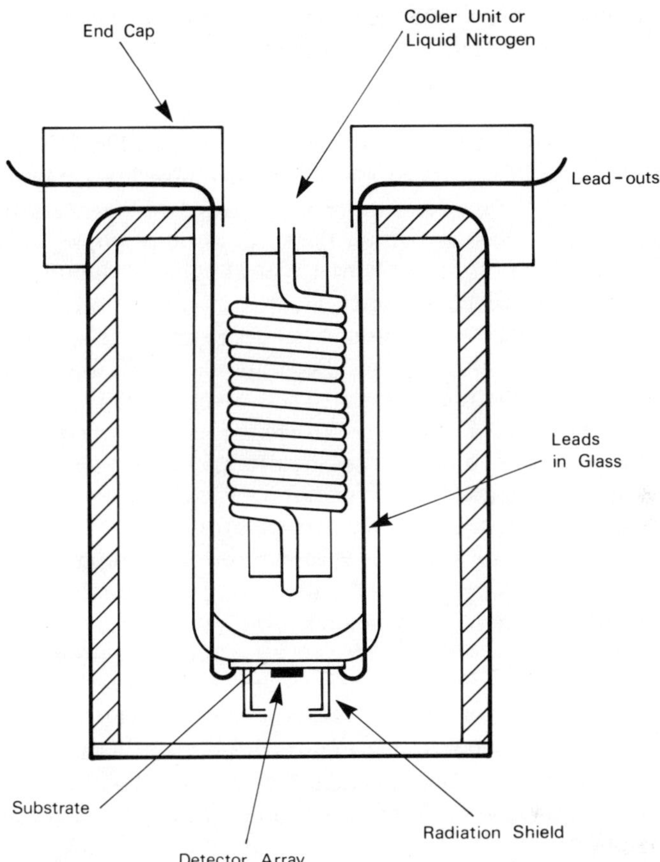

Figure 5.8 Simplified cross-section of dewar with cooler unit mounted over infrared detector.

in the vacuum space at the base of the inner wall of the dewar and looks out through an infrared transmitting window of germanium, silicon, or sapphire. A simplified cross-section of a typical detector assembly is shown in figure 5.8.

For many investigations some means of quantifying the thermal image is essential. Some thermal scanners have adjustable temperature reference sources built into the imaging system and the detector output signal may be compared with that from the reference source at a specific, known temperature. This method has the advantage that the output signal may be calibrated directly in terms of temperature magnitude. However, the manufacture of small stable black body reference sources capable of maintaining a pre-determined temperature in a variable ambient temperature is not easily accomplished. Many scanners rely upon the use of an independent temperature reference which may be located in the field of view; usually the temperature of the standard may be controlled in 1°C steps within an accuracy of better than ± 0.1°C. Reference sources which have matt-black painted faces have an emissivity of 0.97 (7) whereas controlled black body cavities have emissivities of better than 0.99. Some manufacturers calibrate their reference sources in terms of the theoretical black body temperature which would give the same emission within the spectral range appropriate for the detector system. The image display of a typical system is equipped with an isotherm function by which a signal is superimposed upon the grey tone picture so that all surface areas with the same temperature are presented as saturated white. The isotherm can be adjusted as required within the temperature range of the thermogram. The isotherm facility can be used to measure the temperature of an area relative to the temperature of the reference standard: the isotherm can be adjusted first to match the reference and then to match the area of interest. The difference between these two measurements can be used to calculate the unknown temperature.

Various accessories are available with commercial equipment and many thermographic scanners can be used with a computer to store and process data. Maximum, minimum and mean temperatures within an area of interest pre-selected by a tracker ball marker facility is a useful option for medical thermography. Alternatively, two areas can be displayed as squares or rectangles whose height, width and position can be selected independently of each other. The maximum and minimum video levels within each box are displayed on demand. In some systems the scanning lines making up the image can be displayed as temperature profiles across the picture (A scan). Additional options include colour display, alphanumeric facilities, histogram analysis (see Plate I), tape storage and teletype hard copy output compatibility.

In some thermographic cameras, avoidance of picture flicker is achieved by electronic storage techniques. Thermal images of high thermal and spatial resolution may be obtained in this way but such systems are unsuitable for fast dynamic studies. In general, scanners are now designed to function at frame rates greater than 25 frames/s. By using multiple detectors it is possible to achieve an improvement in image quality and still operate at these high frame rates. Lawson (23) has described the principal methods used in scanning systems with multiple detectors; these are illustrated in figures 5.9 and 5.10. In the case of an oscillating scan with a linear array of N elements large enough to cover one dimension of the field of view (θ) with one element per line, the angular resolution (m) will improve as N increases.

A banded scan with a smaller array is illustrated in figure 5.9(b) where the scan is provided by a rotating mirror drum with angled faces. The drum

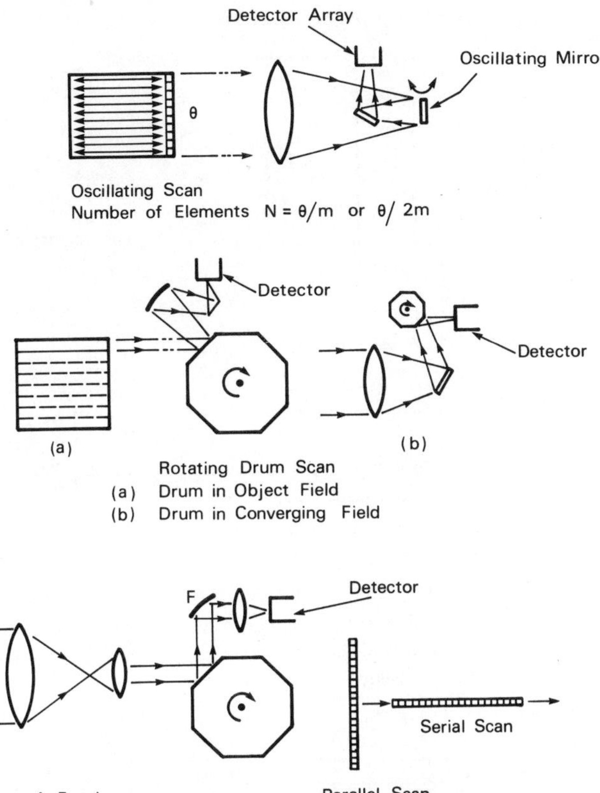

Figure 5.9 Optical elements of some typical infrared scanning systems (23).

may be placed in the object space or in the image space behind a focusing lens. In the former position the drum size becomes too large for higher resolution images but this difficulty can be overcome by using a converging field to permit the use of a smaller drum. In some systems an afocal telescope is used to minimize some of the inherent optical and scanning problems (23).

Most multi-element images employ parallel scanning in which the detector elements of the array scan the scene in parallel. The bandwidth per channel is reduced in proportion to the number of elements and the gain in NETD is proportional to $N^{1/2}$. Because each line in the image is scanned by a different detector, channel uniformity must be high (within a few percent) or else a poor quality picture results.

In serial scanning each detector scans in sequence over every point in the scene. The bandwidth per element is the same as the bandwidth for a single-element detector but the signals add in direct proportion to N, the number of elements, whereas the noise adds in proportion to $N^{1/2}$ as with

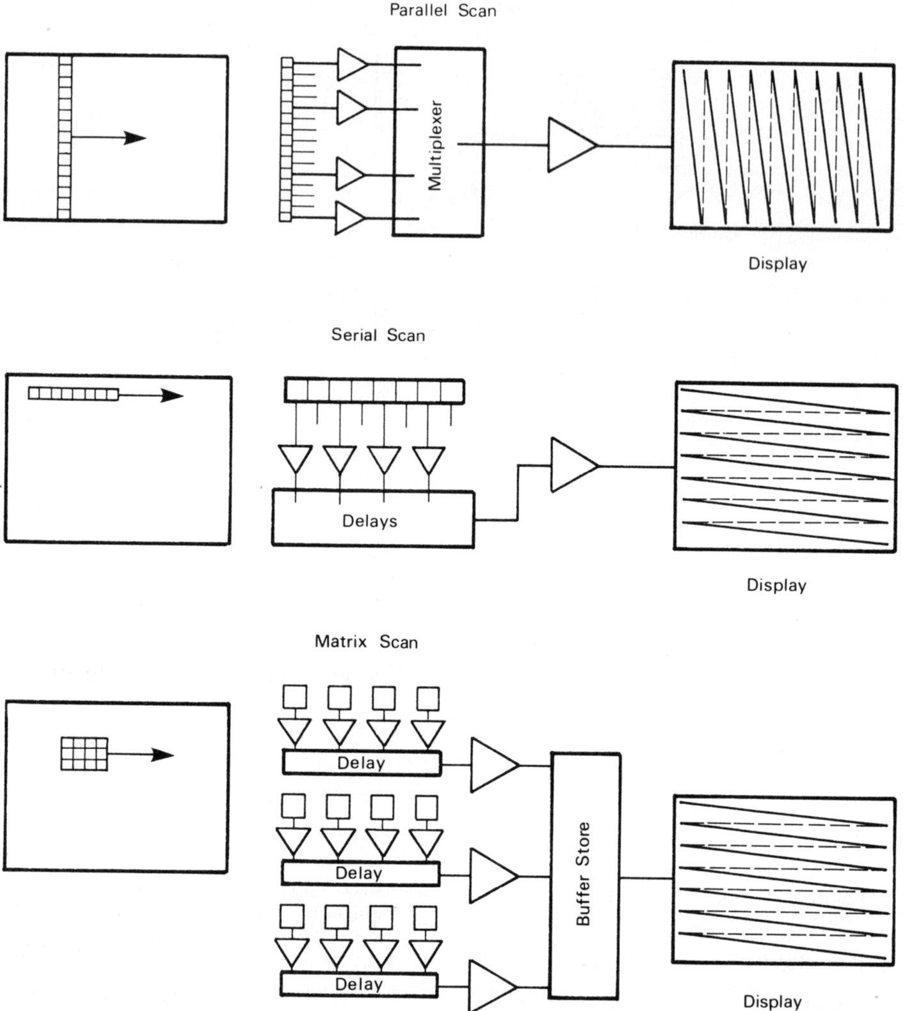

Figure 5.10 Schematic of some typical scan-display interfaces (23).

the parallel scan system. The advantage of serial scanning is that the picture is inherently very uniform without the need for channel matching since all picture points are treated identically. However, high scanning speeds are required to achieve flicker-free pictures and to overcome this disadvantage matrix scans are used. The serial component of the scan reduces the problem of channel matching and the parallel component allows for a reduction in scan speed.

5.5.1 *Mullard SPRITE detector*

This detector is a new kind of CMT detector in which the signal processing takes place in the element (42). In a conventional in-line array used in serial scanning, the signal from each infrared-sensitive element is pre-amplified, delayed and then added to the signal which is generated in the following element. The pre-amplification, delay and summation circuitry is external to the detector and consequently the leads from each element have to be brought out of the detector encapsulation. In the SPRITE detector the line of individual elements is replaced by an infrared sensitive strip mounted on a sapphire substrate (figure 5.11). Whereas each element in a linear array must have two electrical connections and is connected to a pre-amplifier, the SPRITE detector has only three leads so that it considerably reduces the number of leads from the encapsulation and completely dispenses with time delay and summation circuitry. The principle of the new detector may be described in the following way. When a small region of the detector strip is exposed to infrared radiation, excess current carriers are generated which "drift" towards the readout region at a velocity which is determined by the material of which the strip is made and the magnitude of the bias current. The drift velocity is matched to the velocity at which the image is scanned along the strip so the excess carriers are swept along in sequence with the image and all of the carriers arrive at the readout region at the same time.

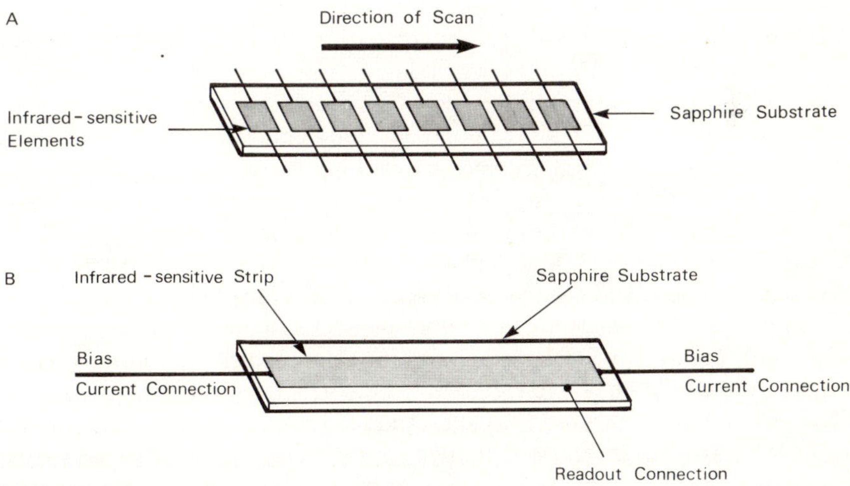

Figure 5.11 Comparison of a linear array of single detectors (*A*) and the Mullard SPRITE detector (*B*).

The SPRITE detector has a detectivity close to the theoretical limit and the performance is therefore controlled by the background radiation. However, since the signal has to be integrated within the detector, the area normalized detecting D^* is greater than the background limited detectivity (D^* BLIP) obtained with conventional detectors. SPRITE detectors are ideally suited for matrix scanning and have been incorporated into high-speed imaging systems.

5.5.2 *Future prospects*

In the past, most infrared imaging systems have had either a single detector or at most a small number of detectors. The development of solid state television cameras makes use of technology which in time could transform infrared imaging cameras. The use of large integrated circuits and large two-dimensional arrays of silicon FET as the basis of TV cameras might be considered as the analogue of a future infrared camera. Putley (40) has described the development of infrared imaging techniques and has described how large detector arrays could be developed using specially doped Si as detector elements. Although the long wavelength cut-off of the normal Si p–n junction is about 1 μm, the response can be extended to 8–13 μm band in doped silicon by employing extrinsic photoconductivity. Such arrays would, however, require cooling to temperatures of 40 K and this would preclude their use for most applications. An alternative possibility would be to fabricate a direct IR analogue of the visible Si array out of a narrow gap semi-conductor such as InSb or CMT. Unfortunately the technology of such materials is far behind that of Si. The requirement for IR devices will continue to be much smaller than that for Si computer circuits, and so it is unlikely that this type of detector will be developed commercially.

Putley suggests that a more likely development will be a hybrid device consisting of a detector array and a matching Si CCD readout chip. The detector array could be connected to the Si chip possibly by a deposition of the infrared material onto the Si chip and then forming the IR detectors *in situ*. Arrays with 10^6 elements should produce a picture quality comparable to that of visible TV but would need an individual element sensitivity of only 10^{-3} of that required for the detector in a single element system. This type of camera would not require any mechanical scanning elements and operation at room temperature would be possible. It should be noted, however, that in uncooled CMT, the fundamental characteristics, such as the effective band gap and the intrinsic carrier concentration, change rapidly with temperature and some form of temperature stabilization would be required for satisfactory operation. Further improvement in sensitivity would be achieved by lowering the

Plate I (A) Liquid crystal 'strip' thermography for measuring breast surface temperature distributions. (B) Colour thermogram showing increased temperature over upper inner aspect of R breast associated with carcinoma; recorded with Aga Thermovision and Integrator Unit. (C) Barr and Stroud IR 11 colour thermogram illustrating histogram display of surface temperature distribution over R breast: temperature magnitude is displayed along abscissa (courtesy Barr and Stroud Ltd). (D) Typical failure thermogram of a 33kV surge divertor. The 'hot' portion was 4.4°C above normal and disintegrated 6 months later (courtesy Aga Thermovision).

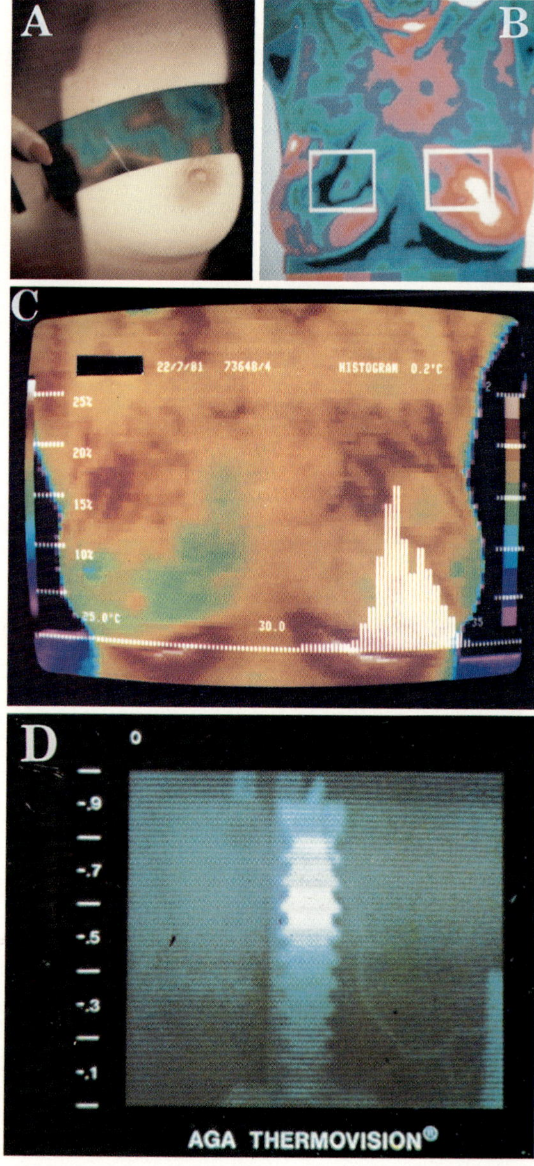
A
B
C
22/7/81 73648/4
HISTOGRAM 0.2°C
25%
20%
15%
10%
25.0°C
30.0
35
D
0
-.9
-.7
-.5
-.3
-.1
AGA THERMOVISION®

operating temperature below room temperature. Pyroelectric materials are cheaper to produce than semi-conductors like CMT and since there is the possibility for depositing the pyroelectrics direct on the silicon it should become possible to produce a pyroelectric-CCD hybrid detector in the near future (43).

Significant improvements in the methods of detector fabrication with consequential improvements in image quality have been made already, but for medical and industrial applications, image processing and data storage are also important. Quantification and analysis of the thermal image are usually necessary and implementation of such processing can increase substantially the overall cost of a thermographic system. Future cameras will be provided with output facilities which will enable data to be stored and subsequently analysed on "in-house" computing equipment.

The history of detector development is represented diagrammatically in figure 5.12 (40).

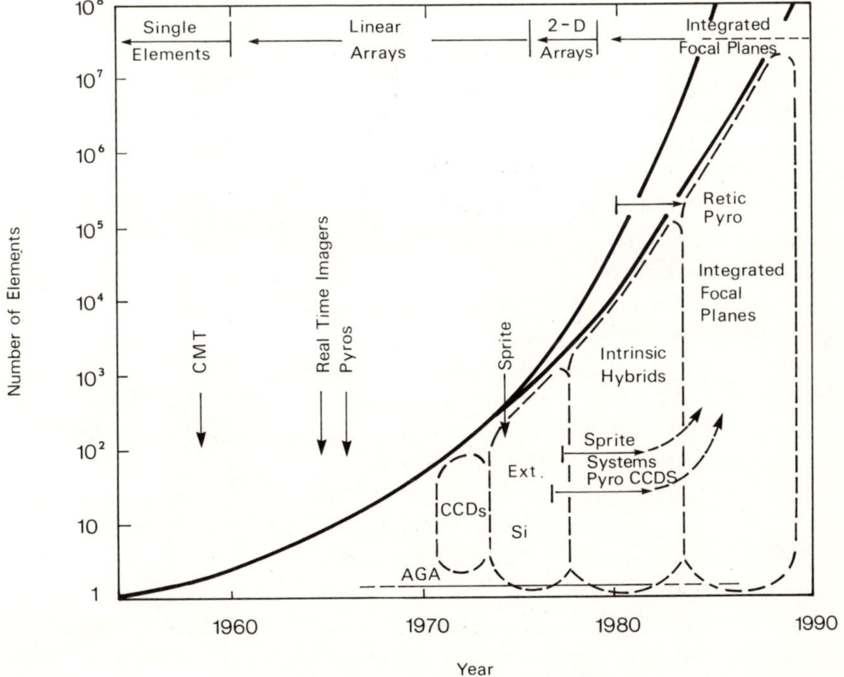

Figure 5.12 Diagrammatic representation of the historical development of infrared detectors (Courtesy Dr E. H. Putley (40)).

5.6 Medical thermography

5.6.1 *Physiological factors*

Thermally, the human body may be regarded as having an "inner core" where heat production is centred and the temperature falls progressively with increasing distance away from this area (44). Whereas deep body temperature is maintained at about 37°C, the temperatures of superficial tissues are influenced by proximity to the core as well as the effects of subcutaneous heat production and the thermal properties of the underlying tissues themselves. The skin of the extremities of the hands and feet is the coolest area.

Although there are small temperature differences within the body, the principal variations in temperature occur over the skin, in the superficial tissues and at the body's extremities. An unclothed, healthy person in equilibrium at 22°C might have a skin temperature which ranges from 35°C over the sternum to 25°C over the feet. The surface temperature distribution is characteristic of the individual and depends upon a large variety of physiological and physical factors. Since skin is the interface between the deeper tissues and the environment, it plays an important role in maintaining body temperature within the fine requisite limits. It must be emphasized that infrared thermography images surface thermal patterns: although these result from the effects of many factors, the absorption of infrared radiation of the wavelengths detected (2–15 μm) is such that transmission through the skin may be considered to be negligible.

Body temperature is controlled within narrow limits by modulating tissue metabolism and the blood flow in skin and subcutaneous tissues. Consequently, the study of the skin temperature distributions can give insight into many physiological problems, particularly those concerned with thermoregulation and metabolism. For this purpose thermal imaging is advantageous. Although skin temperature may be measured with thermistors or thermocouples, these devices are not really suitable for mapping surface temperature distributions over large areas.

In medical thermography the ambient temperature is usually kept about 20°C and consequently the rate of thermal energy flow from skin to the environment is principally the result of convection and radiative transfer: evaporative heat loss is minimal. Blood perfusion plays an important role in modifying skin temperature and the role of thermography as an indicator of this effect has been discussed by Love (45). Venous networks allow blood returning to the vena cava to flow either in deep veins or in the superficial system, or both. The superficial veins course just below the skin, frequently at depths not greater than 2.5 mm and are superficial to the subcutaneous fat layer. The skin temperatures associated with blood flowing through these vessels and the overall temperature distribution over the skin surface will reflect blood perfusion in the tissues, tissue thermal conductivity and metabolic heat generation. These factors are evident in the bio-heat transfer equation which may be used to predict detailed local temperature distributions and thermal energy transport in living tissue (46). The temperature distribution near the surface of the body may be approximated by the equation (47)

$$k\frac{d^2T}{dy^2} + w_b c_b(T_a - T) + q_m = 0$$

where k = thermal conductivity of tissue

w_b = mass perfusion rate of blood per unit volume

c_b = specific heat of blood

T_a = arterial blood temperature

q_m = metabolic heat generation rate per unit volume

y = dimension perpendicular to skin surface

By using appropriate boundary conditions and physical data, Haberman *et al.* (48) used this equation to predict thermographic temperatures resulting from local external heat sources. This type of work and also the work of Love (45) indicate that skin temperatures result primarily from blood perfusion to the tissues and the blood flow in the superficial veins and that metabolic heat generation is a second-order effect. Experimental investigations by Torrell and Nilsson (49) and clinical observations by Jones and Draper (50) support this finding, providing the subject is resting. When thermography is carried out in cool ambient conditions, a blood vessel carrying warm blood can affect the skin temperature providing it lies within about 6 mm of the skin surface. Consequently the thermal effect of blood vessels may be evident even though the vessels themselves are not visible. Nevertheless, there are situations where metabolic heat affects the thermal image significantly. Energy expenditure or conservation and concomitant metabolic heat generation in exercise and in extremes of environmental temperature (44) are of particular interest to physiologists. Considerable attention is also being paid to the problems of heat generation in the neonate (51) and the elderly (52). Neither the very young nor the aged have efficient mechanisms of thermogenesis. It has been suggested that brown adipose tissue over the nape of the neck and interscapular region of neonates acts as a source of heat production by means of "non-shivering thermogenesis". Thermography has been used to study the distribution of heat emission from brown fat in the neonate (53) and in man (54). There is also evidence that some large rapidly growing malignant tumours generate significant metabolic heat (55). In general, however, medical thermography is dependent largely on changes in blood perfusion in tissue or blood flow in superficial vessels.

5.6.2 *Preparation of the patient*

Clinical thermography is carried out best in a draught-free, cool environment in a constant ambient temperature of about 20°C. At this temperature it is easier to achieve standardization of the imaging technique. The skin of the patient should be cooled uniformly for a period before the thermographic examination starts, so that the surface thermal pattern is accentuated and remains constant throughout the investigation. Complete equilibration might take 40 minutes or so, but in most instances the thermal pattern is relatively stable after about 15 minutes. The examination room should be air-conditioned if reliable quantitative measurements are to be obtained. The thermostat controlling the cooler-heater unit must be sited appropriately within the examination area and the temperature monitored continuously. It is not necessary to paint the walls of the room black but care must be taken to avoid the reflection of infrared by highly polished metal surfaces into the field of view of the camera (56).

Before examination the patient disrobes to expose the area to be examined. It is necessary for the patient to remove tight-fitting clothing which might restrict blood circulating through the area being investigated. Passive cooling of the patient for about 15 minutes removes excess heat from the body surface, sharpens up the thermal pattern and stabilizes the temperature distribution over the skin and subcutaneous tissues. It is important that the patient should sit or lie restfully throughout this period without anxiety; the patient should not be overcooled or uncomfortable during this period of preparation. Unless otherwise indicated, patients should be advised not to eat, drink or smoke immediately before a thermographic examination. Changes in temperature can be expected after heavy meals or hot drinks and the vaso-constrictive effects of nicotine on the peripheral vasculature can produce misleading results. The skin surface should be clean and free from talcum powder or cosmetics and patients should be asked to avoid rubbing or scratching the surface to be examined. When investigations are carried out in a ward or clinic which is not air conditioned, local cooling of the skin surface can be achieved with a hand-held fan or by spraying the area with spirit which, upon evaporation, cools the underlying surface. Both methods are effective for localized cooling but must be used carefully if thermal artefacts are to be avoided.

5.6.3 *Clinical techniques*

Most important among the clinical techniques are:

(1) screening for occult malignant disease—although this method has been used extensively for the early detection of breast cancer (57, 58), and in the differential diagnosis of pigmented skin lesions (59), thermography lacks sensitivity and specificity (60);
(2) delineation of the extent of suspected or known disease (61);
(3) identification of areas with abnormal temperatures which might be the cause of functional impairment of underlying organs or glands—for example, male infertility is sometimes associated with an abnormal temperature increase of the testes which may be detected by scrotal thermography (62, 63);
(4) monitoring the effects of various forms of therapy such as reconstructive surgery, radiotherapy, hyperthermia or treatment with hormones or drugs (64, 65, 66);
(5) for assessing the prognosis of certain diseases; tumour metabolism and malignity can be reflected by thermal changes overlying the tumour (59, 67);
(6) identifying functional deficiencies and vascular disorders (68, 69);
(7) studying the effects of chronic or acute trauma; these investigations range from the thermographic examination of "psychogenic pain" to the assessment of burns and frostbite (70, 71, 72);
(8) physiological research such as energy metabolism and peripheral vascular investigations (54, 73).

The ease with which thermography can be used has resulted in a large number of investigations, many of which are of small value because inadequate care has been given to patient preparation, thermal calibration and clinical assessment. The subject has a large bibliography (74, 75, 76, 77, 78, 79) and it is beyond the scope of this chapter to describe how the technique can be used in all clinical situations, but attention is drawn to a select number of applications which illustrate the various practical aspects and principles of the technique.

Vascular disorders. Since the regulation of body heat is largely dependent on blood flow through peripheral vessels, it is not surprising that vascular disease affects the temperature of body extremities.

Raynaud's disease is a vascular disturbance caused by spastic contraction of the smaller arteries of the extremities, particularly in a cold environment. Thermographic investigations are helpful in the diagnosis and assessment of this disease. Symptoms can be induced by immersion of the limbs in iced water and thermography may be used to examine the disturbance under various conditions before and after medication (73).

Thermographic techniques are being used to assess the effects of peripheral arterial disease on tissue temperatures. Peripheral arterial disease can cause ischaemia necessitating amputation of an affected limb. Selection of the optimum level of amputation in these patients is one of the key factors which determines successful rehabilitation. Most surgeons depend upon clinical assessment of tissue viability and also upon the measurement of physical parameters such as skin blood flow, perfusion pressure and ultrasonically derived arterial pressure gradients. Thermography has also been shown to be an accurate and reliable method of determining the best site of amputation. In a controlled thermal environment gradients in skin temperature of extremities are related primarily to the heat supplied from the body core by blood convection and conduction. In an affected limb a hyperthermia pattern depends on the presence of collateral pathways when the main arterial branches are occluded whereas a hypothermal pattern depends on the existence of an arterial stenosis with a decrease in blood flow. The most important factors in the thermographic assessment have been found to be the severity of the longitudinal thermal gradient along the limb and the presence or absence of local hypothermia (80). Thermograms and thermal profiles along the length of the limb are useful for assessing the severity and examples are illustrated in figure 5.13. The non-invasive nature of the technique allows the method to be used even on patients who are in extreme pain.

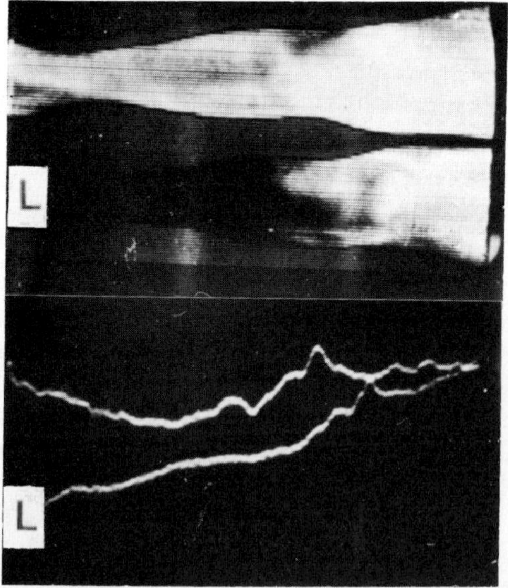

Figure 5.13 Thermogram and '*A*' scan thermal profiles of a patient with ischaemia of left leg (EMI Thermal Scanner; black—cold, white—hot) (Courtesy Dr V. A. Spence).

Although the aetiology of varicose veins might be in doubt it is well established that incompetence of the venous valves is the basic underlying defect. Incompetent perforating veins in the leg may be found thermographically (68). The technique is interesting because the pattern changes are induced physically and the way in which the temperature distribution alters under controlled stimuli is used to assess valvular incompetence. With the patient supine the veins are drained by raising the leg to an angle of 30 to 40 degrees for one minute, the leg temperature being lowered with an ice-cooled wet towel and an electric fan for five minutes. By applying a tourniquet round the upper third of the thigh occlusion of the superficial veins can be achieved and the patient then stands and exercises the leg. Areas of "rapid rewarming" below the level of the tourniquet indicate the sites of incompetent perforating veins. The test is simple to perform, and atraumatic; it gives an objective assessment which is more precise than subjective clinical methods.

Deep vein thrombosis. A number of independent investigations have shown that thermography is a reliable non-invasive technique for the diagnosis of recent deep vein thrombosis. This vascular disorder complicates major surgery and illnesses such as myocardial infarction and stroke. Confinement, oestrogen therapy and oral contraception also carry some risk. The clinical signs of DVT are calf pain and tenderness, oedema and induration, and an increase in limb temperature, but these are not always reliable. Phlebography, radioactive iodine fibrinogen uptake, transcutaneous ultrasound, impedance and strain gauge plethysmography can all be used as tests for DVT. Thermography is, however, the only technique available for the large number of patients who present with symptoms and signs referrable to the legs and for which no preparation of the patient other than a cooling period is required (69). Whereas phlebography is the most accurate means of detecting the presence of a deep venous thrombus and has a high sensitivity and specificity, it is time-consuming and requires skilled medical and radiological personnel.

Thermography has been used to identify DVT by observing delayed cooling of the affected limb. In the absence of DVT, the thermograms of a resting healthy subject show the thighs and calves to be cool. In the calf, the subcutaneous border of the tibia and patella are cooler than the surrounding muscle. Recent calf vein thrombosis produces a diffuse increase in temperature which may involve the whole or greater part of the calf and temperature increases of about 2°C are not unusual. These thermal changes are evident on both supine and prone views. The thermogram remains positive during thrombus propagation and for some days afterwards. This thermal activity is probably caused by the local release of vasoactive chemicals associated with the formation of the venous thrombus which cause an increase in the resting blood flow. The thermal consequence of this blood flow difference between the two limbs is accentuated by examination of the patient in a cool ambient and by the relatively cool temperature of the normal limb.

Although thermography is able to predict extensive DVT with a high level of accuracy, the presence of venous insufficiency, extensive inflammation or other heat producing lesions may obscure the recognition of concomitant DVT. Nevertheless, the innocuous nature of the test enables it to be used daily if required. Although thermography is less accurate than phlebography for the diagnosis of early limb thrombosis, there is no doubt about its value if careful attention is paid to the underlying anatomy of the limb (82).

Scrotal thermography. In man the testes are situated in the scrotum apparently to maintain their temperature a few °C lower than core body temperature. If this temperature differential is abolished, spermatogenesis is depressed and may eventually cease. The temperature of the testes is reflected by the surface temperature of the scrotum which after equilibration in an ambient of 19°C is typically 30 ± 1°C. The average temperature difference between both testes is less than 0.5°C. A varicocele caused by reversal of blood flow in the internal spermatic vein can interfere with normal circulation of blood in the scrotum and can produce an increase in the temperature of the testes, the magnitude and extent of which depend upon the size of the varicocele. This condition is not uncommon and affects about 10 percent of young men and about 20 percent of subfertile males who have a variable degree of impaired spermatogenesis. It has been shown that thermographic abnormalities correlate well with the presence of internal spermatic vein reflux demonstrated by retrograde caval venography (62, 63). Because thermography is non-invasive, it is preferable to phlebography for routine investigation; it involves no radiation hazard so it may be repeated as often as necessary in sequential studies on an individual patient before and after treatment.

Scrotal thermography is used in the following situations:

(1) to determine the thermal effect and extent of clinical varicocele,
(2) to investigate infertile or subfertile men who might have an unsuspected varicocele,
(3) to examine patients who have had corrective surgery and to determine whether any residual veins are of significance after ligation of the varicocele.

A temperature greater than 32°C is considered to be abnormal especially if this is associated with thermal asymmetry greater than 1°C(83). Figure 5.14 shows a typical abnormal thermogram. The technique is reliable and easy to perform. Thermographic examination of the front underside and posterior of the scrotum is necessary. This may be accomplished by an

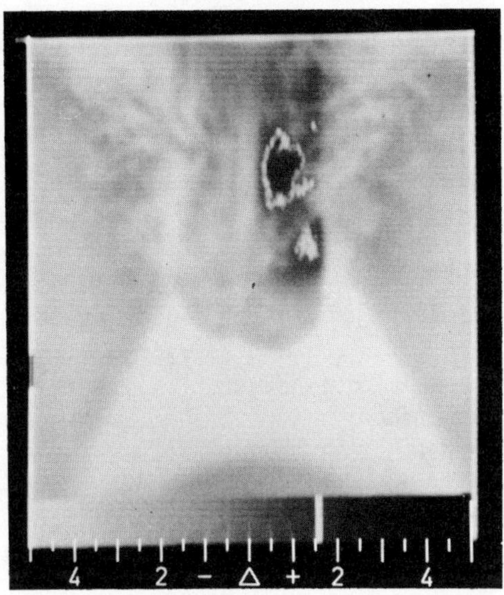

Figure 5.14 Scrotal thermogram showing thermal asymmetry associated with left varicocele (R = 30°C; L = 33.5°C) recorded with Aga Thermovision.

anterior view and an underview achieved with the aid of a highly polished aluminium or front-surface-silvered mirror positioned so as to reflect the thermal image of the underside of the scrotum. Alternatively the patient may be examined in the knee-elbow position and examined posteriorly. Some workers have found that the Valsalva manoeuvre increases the sensitivity of thermography specially when the varicocele is small, or sub-clinical (84).

Female breast thermography. Although the temperature distribution of the surface of the female breast varies considerably from one person to another, each healthy individual has a characteristic pattern which remains basically unchanged over a long period of time (85). Changes in deep body temperature such as those associated with diurnal variations and monthly cyclical variations affect surface temperature levels. For example, an increase in surface temperature of the breast of about 1°C and intensification of the thermal pattern prior to menstruation have been found to occur in some women; oral contraceptives can cause temperature elevations, as well as pregnancy and lactation. However, the overall thermal pattern is closely related to the subcutaneous network of blood vessels and the temperature of blood flowing through these vessels.

Benign and malignant tumours can affect both the vascular pattern and the blood flow in the affected breast which consequently alters the temperature distribution and often results in thermal asymmetry between breasts. Lipomas, galactomas and haematomas do not usually cause a temperature elevation and are sometimes associated with a local reduction in temperature. Benign conditions such as fibroadenosis, cystic hyperplasia, fibroadenomata and cysts can cause thermal patterns and temperature changes similar to those associated with malignancy but only in about 20–30 percent of patients (86). Inflammatory conditions such as abscesses and infections cause large local temperature increments. In the case of malignant tumours the surface overlying the tumour is frequently between 1 and 4°C warmer than the surrounding skin and it was this observation by Lawson in 1957 (87) that led to the development of breast thermography. The subject has a large bibliography, and although thermography is used infrequently in the United Kingdom, it is used extensively in some European countries. The technique is of interest historically (it was the first thermographic test to be used clinically) and also physiologically. A large amount of information has been obtained about the temperature distributions in healthy and abnormal breasts which in turn has provided a greater understanding of breast tumour physiology (88).

In an ambient temperature of 19–20°C after an equilibration period of 10–15 minutes the breast surface temperature of a healthy woman ranges within 28 and 36.5°C. Vessels which course near the surface of the breast carrying blood from deep within the breast result in prominent thermal profiles over the skin surface with widths at half height of 1–3 cm with gradients of between 1 and 2.5°C cm^{-1} causing temperature increases of up to 4°C (89). There are at least three types of pattern:

(1) patterns with minimal thermal markings; these are associated with cold avascular breasts with a mean breast temperature of about 30 ± 2°C,
(2) patterns with prominent thermal markings over both breasts with a mean breast temperature of about 32 ± 2°C and
(3) "path-type" thermal patterns (sometimes called "leopard pattern") which cover both breasts and often spread over other areas of the body.

There are three areas over which the thermal pattern is frequently pronounced. These are (1) the upper outer aspect of breast over the lateral thoracic artery which appears as an area of focal thermal activity, (2) an elongated area radiating medially from the areola overlying the subcutaneous mammary vein; and (3) an area circumscribing the areola where subcutaneous blood vessels form deep and superficial plexuses.

The development of a malignant tumour within the breast usually changes the temperature distribution over the surface of the affected breast and this often occurs in these well vascularized areas. The thermographic features considered to be abnormal are: (1) a localized area of temperature increase of about 1.5°C or more; (2) a localized increase in vascularity and its associated thermal pattern (this might be in the form of venous engorgement, dilatation or an increase in the number of superficial blood vessels); (3) unilateral increase in temperature over the areolar area; and (4) a generalized increase in temperature of one breast (90). Plate IB and figure 5.15 show thermograms of patients with prominent vascular patterns associated with carcinoma. Although precise thermal symmetry of both breasts is rare, most women who do not have breast nodularity or a localized lump tend to have similar surface thermal patterns over both breasts. (The essential procedure is to compare both breasts using one as a control for the other). Experience at the Royal Marsden Hospital, London, where 20 000 women have been examined clinically and thermographically, suggests the following conclusions (86):

1. When both breasts are asymptomatic, the occurrence of significant thermal asymmetry (typically greater than 1.5°C) is less than 5 per cent.
2. When there is a lump (or localized nodularity) which is of no clinical or

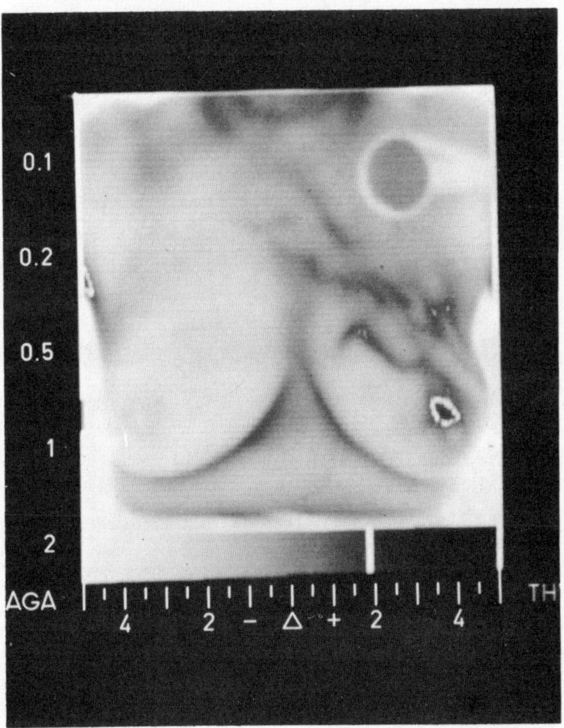

Figure 5.15 Thermogram of patient with carcinoma, left breast, showing 3°C asymmetry recorded with Aga Thermovision.

mammographic significance, about 15–20 per cent of these patients have an abnormal thermogram.
3. Twenty to thirty per cent of lumps that are palpable and on the basis of clinical or mammographic examinations require surgery, cause a thermal abnormality.
4. Sixty to seventy percent of Stages I and II cancers have abnormal thermographic features and eighty to ninety percent of Stages III and IV cancers.

The temperature increase in neoplasia is associated with changes in superficial vasculature and in larger tumours with tumour metabolism. Theoretical calculations show that 2 mm diameter vessels carrying blood at 37°C can affect significantly the surface temperature of the breast providing the blood vessel is not at a depth greater than 6 mm (49). Thermography largely depends on the effect of the neoplasm upon such vessels (see figure 5.15).

Infrared thermography lacks the sensitivity and specificity to be used alone as a diagnostic procedure. Nevertheless it is a useful adjunct to clinical examination and mammography for selected patients. It is completely safe and because the thermal pattern of a healthy individual remains unaltered for many years it can be used at frequent intervals to monitor patients with chronic benign disease, and for identifying and screening "high risk groups" (91). The technique has been used to screen asymptomatic women for malignant disease (58); to assist with the diagnosis of palpable disease (92); to monitor patients receiving radiotherapy (93), and to monitor the efficacy of treatment by chemotherapy and hormones (66). Although thermography misses at least thirty percent of very early cancers, there is some evidence that these patients have a better prognosis than those with the same stage of disease who have abnormal thermal patterns.

Liquid crystal plate and sheet thermography have been used in place of infrared imaging by some investigators. Plate IA illustrates the use of a colour-temperature sensitive sheet to measure surface temperature distributions. In the case of plate thermography it is not always possible to obtain uniform contact between the plate and the part of the breast being examined: the shape of the breast and chest wall influence the area that can be investigated. However, the method has been used extensively by Tricoire *et al.* (30) and Tonegutti *et al.* (35).

Thermographic examination of arthritis and similar diseases. Arthritis is frequently a chronic inflammatory lesion and since heat is one of the classical symptoms of inflammation, thermography has proved to be a useful method of demonstrating localized inflammation and evaluating treatment (94, 95). Pain, which is a dominant factor, may be controlled by analgesic drugs, but subsequent functional tests will not necessarily reflect the change in the patient's inflammatory state. Quantitative thermography provides a means of following minor changes in joint temperature and of monitoring the effects of drug and physical therapy. To this end Collins *et al.* (94) devised a quantitative system using a Thermographic Index (T.I.). The concept was based on the finding that under given conditions normal peripheral joints could be cooled to a predictable temperature range. Two temperature ranges were chosen: 26–32°C for lower limbs and 28–34°C for the joints of the upper limbs. The thermographic index is calculated from the expression:

$$\mathrm{TI} = \frac{\sum \Delta T a}{A}$$

where ΔT is the difference in °C from the base temperature of either 26°C or 28°C to each isotherm temperature recorded, a is the area of each isotherm and A is the total area of the thermogram (in cm^2). In practice the area of each isotherm within the selected area is measured and multiplied by its baseline difference. Using 0.5°C intervals each total is added together and divided by the total area to be quantitated. This formula provides a scale from 1.00 to 6.00. Normal values for a healthy subject are usually less than 2.5 and inflammatory joints are raised to 6.00 (95).

Whereas the procedure for obtaining the required isotherms is straightforward, the generation of the TI from the isotherms is tedious and it is usual to overcome this difficulty by computer processing of the thermal image and the selected area of interest. Preparation of the patient and standardization of the technique are important factors. Reproducible results are achieved when examinations are carried out in a uniform constant ambient temperature and preferably at the same time of day to avoid diurnal variations. The Thermographic Index has been used to assess the effect of anti-inflammatory drugs by measuring the rate of change of the index by the test drug, and to measure the anti-inflammatory response to intra-articular steroid injections. The long-term changes in TI often occur before observable changes in the clinical condition of the patient.

It may be noted that similar quantitative measurements can be achieved with integrator-type devices supplied by some thermographic equipment manufacturers.

Assessment of trauma. The treatment of a burn depends upon the depth of the penetrating injury and the extent on the skin surface. A first-degree burn shows an erythema on the skin, sometimes accompanied by blisters. A third-degree burn is deeper, shows a complete absence of circulation and the skin, which is white in colour, is insensitive to pain. The classification is, however, not always clear-cut and infrared temperature measurement is a useful objective method of assessment (71, 96). Third-degree burns are found to be about 3°C colder than the surrounding normal skin. The speed with which the examination can be made and the absence of physical contact between patient and thermal scanner are important clinical advantages. Areas that display temperatures of 2–3° C below unburned skin temperature require grafting whereas other areas displaying temperatures 1°C below to 3°C above unburned skin temperature tend to heal without grafting.

Pressure sores caused by long periods in bed and bruising prior to hospitalization have been monitored and assessed thermographically (20). Problems associated with stress to insensitive tissue such as those arising from limb prostheses are difficult to resolve and early detection is important if prevention and control are to be achieved. Thermography has been used to aid evaluation of prosthetic devices and to detect irritated tissue prior to frank breakdown (97).

Thermal imaging of sports injuries and bone fractures indicates areas of hyperthermia (37). Of more significance is its use for choosing the surgical approach in complications of compound fractures of the leg (98). There is some indication that thermography identifies areas of good vascularization which is indispensable for the healing of the surgical scar so that subsequent skin damage and necrosis are reduced.

Evaluation of spinal root compression syndromes. Thermography is being used to show evidence of nerve root irritation. It is claimed that thermal studies correlate well with clinical and surgical findings for root compression syndromes at low cervical and low lumbarsacral levels, particularly when the extremity dermatomes are included (99, 100). A normal thermogram of the spine is characterized by a central zone of either increased or decreased heat in the region of the spinal processes from the cervical spine down to the lower lumbarsacral spine. The thermal pattern is basically symmetrical: even the sacroiliac joints may show symmetrical localized heat emission. The surface topography of the patient is important and determines to a large extent whether or not the linear thermal pattern over the spinal processes is relatively hot or cold; when the body contour results in a 'gulley' running down the patient's back, the surface temperature over the spinal column is relatively warm, whereas a spinal column that is protuberant is relatively cold.

An abnormal thermogram can provide evidence of asymmetric heat production at dermatomes and myotomes. It has been suggested that temperature changes may be related to reflex sympathetic vasoconstriction within affected extremity dermatomes and metabolic changes or muscular spasm in corresponding paraspinal myotomes. Hypothermia (about 1°C) usually occurs at the levels of the affected extremity dermatomes but hyperthermia can also occur particularly in the case of hands and feet. The precise mechanisms of thermal changes are not understood completely but thermography appears to be a useful complement to myelography in identifying clinically significant abnormalities.

The thermographic study of pain, its anatomic location in relation to its physical cause and the objective evaluation of this subjective condition is an interesting and important subject. Organic disorders such as sympathetic dysfunction can be misdiagnosed as psychogenic pain. Although a number of methods have been developed to assess sympathetic nerve functions such as the electrical skin resistance tests, calorimetry, plethysmography and thermocouple studies, thermography allows large surface areas to be examined without physical contact and provides data for evaluation. This latter aspect is important because it allows temporal changes to be observed before and after treatment and in the case of injury thermal assessment provides objective quantitative data (70).

5.6.4 *Infrared thermography and hyperthermia*

The treatment of superficial cancer by localized hyperthermia requires careful control of the temperature within the tumour and in overlying healthy tissues. The technique should generate adequate heating (usually in the range 42–43°C for periods of about 45 minutes) throughout the tumour without causing similar temperature increases in surrounding normal tissues. To achieve this end it is necessary to optimize the heating technique with respect to the tumour being treated and to monitor the surface heating during hyperthermia therapy. Thermography is useful (101) for:

1. Selecting the most appropriate form of heating technique. Heating techniques utilize electromagnetic fields, microwaves and ultrasound to transfer energy to tissues. The thermal characteristics associated with particular applicators and probes may be assessed thermographically before clinical use.
2. Examination of the skin surface overlying the tumour prior to treatment. Tumour vasculature and the vasculature between tumour and heating electrode play an important role in controlling the temperature differential attainable between tumour and normal tissues. Thermography is of help in identifying the presence of any non-uniform temperature distribution in the superficial tissues.
3. Assessment of the skin sensitivity to heat and the effect of vasculature on thermal load. Achievement of therapeutic temperatures depends upon the ability of the patient to withstand temperatures of 42°C or more. The rate of loss of heat from the heated area affects the temperature that can be

attained. This, and also maximum surface temperature that a patient can withstand without discomfort may be assessed thermographically with the aid of a specially designed temperature reference block placed on the patient's skin prior to treatment.

4. Monitoring surface heating patterns during hyperthermia therapy.
5. The measurement of post-therapy changes.

5.7 Veterinary applications

Both infrared thermography and liquid crystal sheet thermography can be used to determine surface temperatures of animals. Infrared techniques are particularly advantageous because they can be applied to unrestrained and agitated animals and may be used in situations where the proximity of an observer might either disturb the animal or involve danger to the observer. Contact thermography on the other hand requires co-operation of the animal being investigated: liquid crystal sheets placed in close contact with the animal's fur or skin are useful for monitoring surface temperatures particularly over limbs and joints after injury. Lelik *et al.* (38) have used the method successfully to monitor the treatment of locomotive disorders in horses, cattle and dogs. The range of surface temperatures they report, from 24 to 38°C for horses and 28 to 40°C for dogs, is wider than that likely to be found in humans: this reflects the higher body temperature of most mammals and the effect of coat insulation.

The influence of the coat of animals on infrared thermography has been considered by Cena and Clark (102). The coat will attenuate radiation originating at the skin; this will be dependent on its thickness and composition. Cena and Clark (103) show that the temperature drop across the insulating layer will be proportional to its thermal resistance. The temperature measured by thermography will be reduced by the ratio $r = H_e/(H_e + H_c)$ where, H_e is the external resistance to heat transfer outside the coat and H_c is the thermal resistance of the coat. For a typical short hair coat other than a sheep's, r would be less than 0.5. For sheep, considerable variations can occur depending on whether the sheep is shorn: for a shorn animal r is close to 1 whereas for an unshorn sheep r approaches zero. Variations in coat thickness can produce apparent thermal changes similar to those caused by illness.

Successful application of infrared thermography in the diagnosis of orthopaedic lesions in racehorses has been described by Strömberg (104). He found detectable increases in surface temperature associated with local elevations of metabolic rate within the lesions and was able to monitor the progress of healing. The monitoring of the reactions of animals to pharmaceuticals during the course of drug testing has also been investigated.

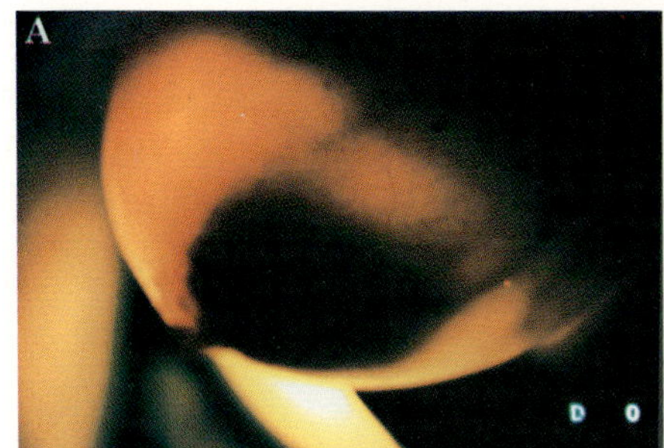
A
D 0
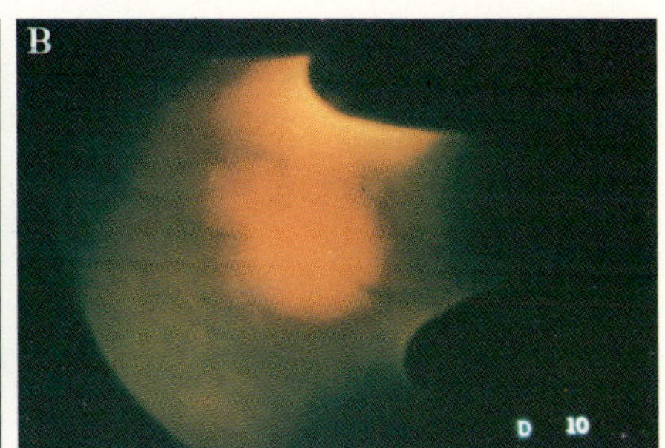
B
D 10
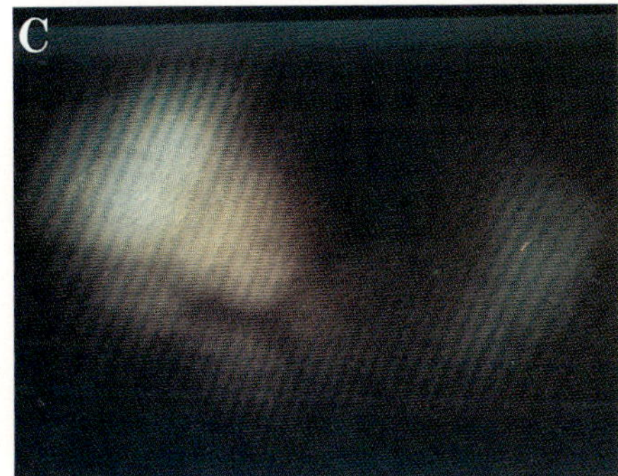
C

D

Plate II (A) A marked opacity in the breast typical of the appearance of carcinoma or bleeding following biopsy. (B) An area of increased brightness typical of the appearance of a cystic lesion. (C), (D) Pseudo-colour images of an opacity in the breast situated close to the chest wall.

5.8 Environmental studies

Most environmental studies are made with scanning systems that are sensitive to radiation in discrete regions of the spectrum. Regions of special interest include the visible spectrum, the near infrared: 0.7—1 μm, the 3—5 μm and the 8—14 μm regions. Cameras have been developed with multiple detectors to image at these wavelengths either simultaneously or sequentially. Nichols and Lamar (105) developed an instrument that scans a scene simultaneously in three spectral regions (0.5—1.0 μm, 3.0—5.5 μm and 8.0—14 μm) and produces an image of the scene as a colour photograph. Si, InSb and GeHg are used as the respective detectors and as the scene is scanned the detector outputs are used to modulate three separate light sources. Each light source is filtered so that it provides one of the primary colours, blue, green or red. The resultant pictures are a result of both reflected and self-emitted radiation from objects in the scene: the colour of objects in a picture indicates their temperature and also their reflective and emissive properties.

Aerial and satellite surveys are being used to study terrain texture, geological structure, the global distribution of geothermal areas and the heat capacity of soils (106, 107). The accumulation of satellite data is more complex than earth-bound imaging but similar spectral bands are often employed: the NoAA-7 polar orbiting satellite is equipped with the advanced very high resolution radiometer (AVHRR) with the following detector spectral bands: 0.58—0.68 μm, 0.725—1.1 μm, 3.55—3.93 μm and 10.5—11.3 μm (108). As part of a $2\frac{1}{2}$ year NASA Heat Capacity Mapping Mission (HCMM) satellite data has been accumulated with 6000 data passes and 26 500 frames of imagery. The programme has confirmed that such data is an effective means of gaining insight into understanding earth resource problems. The mission objectives are defined to be: discrimination of rock types and mineral resources; measurement of plant canopy temperatures to determine the transpiration of water and plant stress; the measurement of soil moisture effects; mapping thermal effluents, both natural and manmade; and coverage of snow fields for water run-off prediction. Achievements to date have been summarized (108) and include information on such diverse topics as the discrimination of rock type in spite of vegetative cover, the identification of frost-prone or frost-free areas, location of shallow water tables, the mapping of industrial pollution and natural thermal phenomena.

There are two useful atmospheric windows in the infrared spectrum: these occur from 3 to 5 μm, and from 8 to 14 μm. Fortuitously, indium antimonide and cadmium mercury telluride cover these two regions. Environmental studies are often carried out from aerial platforms, helicopters, and low-flying aircraft. Thermographic cameras and line scanners may be used to image large areas of the world's surface. In the case of a line scanner the thermal imaging system scans along a line perpendicular to the direction of flight so as to build up an aerial view.

Lechi (109) has described some of the ways in which thermography can be used for environmental quality control. Imaging of sources in which the emissivity is constant gives thermal data which can be understood in geo-physical terms: for example, since the emissivity of non-polluted water is constant (~ 0.97), the surface temperature of a river delta can be interpreted in terms of the dynamic behaviour of the water—where the temperature is higher, the water velocity might be expected to be lower. By studies of this kind, evaluation of the volume of power plant or chemical effluent being discharged into the sea is possible.

Additional information may be obtained by imaging sequentially at 3—5 μm and 8—14μm wavelength bands. The difference between detector signals can be computed: the reflection of solar energy by a surface will be different for these two bands and it becomes possible to evaluate the reflection behaviour in the 3—5 μm band.

Thermal surveys are particularly important for mapping geothermal and volcanic areas and investigating the heat capacity of soils (110). Data can be processed to show not only the temperature distribution over an area of interest (such as the crater of a volcano) but may be used to determine and display the rate at which the temperature pattern varies both spatially and temporally.

Plants, vegetation and trees have been investigated extensively. The radiant power from plant life is closely linked with biosynthesis and disease processes affect the infrared energy emitted by a plant or tree. Plants in stress situations manifest anomalous thermal behaviour and thermography may be used as an indirect monitor of air pollution by its effects on vegetation (106, 107, 109, 111).

In the midst of dense smoke which limits visibility, forest fires can be monitored thermographically; this is a means of searching for "sleeping" fires in forest areas. Small portable thermal imaging cameras are used by fire fighters to find their way through dense smoke; such equipment is now available to local fire fighting authorities (23).

5.9 Industrial thermography

The versatility of thermography, together with the requirements of energy savings, makes it a most cost-effective inspection technique. As process plants become larger and more expensive to replace and operate, less spare plant becomes available. Moreover, technical operations are often close to the design limit and consequently considerable pressure is put on inspection and maintenance departments. Whereas in medical thermography there are relatively few situations in which cost-effectiveness can be ascribed with confidence to a particular test, in industrial problems the situation can be assessed more rigorously. Preventive maintenance can save the user millions of pounds and thermography plays an important role in many of these programmes.

Thermography is being used to assist energy conservation engineers to assess heat losses in industrial plants and to find areas where energy conservation can be made. A principal advantage of the method is that thermal leakage may be located without dismantling insulation. Energy savings have been reported in such diverse industries as cement, paper, forging, foundry, copper, heat treating, refractories, rubber and heavy manufacturing (112). The potential for identification of flaws in insulation, malfunctioning electrical components, and serious heat losses by other causes makes infrared survey a powerful non-destructive tool (113). The technique is well suited for processes involving heat generation, transmission or retention, or where processes are carried out on a continuous basis in refractory-lined units (114); it is widely used by the electricity supply industry where fault identification can be hazardous because of the high voltages produced and problems of access and where faults may be present randomly over an extensive area (115). Thermal images may be recorded at safe distances from structures which is of great importance when high temperatures, dangerous gases or fumes are present or when radioactivity precludes direct viewing (116). Thermography can be used in the following ways.

1. To provide information which will facilitate optimal design and efficient use of industrial equipment and processes.
2. To obtain data about the basic operating characteristics of a working plant, to assess its efficiency, and predict its reliability; temperature monitoring is a well established method of assessing plant condition—its magnitude and its distribution are indicators of variance from normal or acceptable operational performance; a "thermographic reference" of the equipment or machinery obtained under normal operational conditions is a useful guide for future inspections.
3. To locate and measure thermal irregularities so that they may be assessed and that appropriate action can be taken.
4. To check the effectiveness of maintenance and repair work.
5. To adjust or regulate the operation of an industrial process; in this case, thermography may be considered to be an integral part of process and the output of the infrared equipment is used to control the process.

In many industrial procedures it is necessary to control the individual processes within narrow thermal limits even though external factors such as the environmental temperature and humidity might vary considerably.

Thermography allows monitoring to be made under different conditions so that unnecessary component failure or product damage can be avoided (116).

A thermogram of a surface indicates only the relative temperature distribution: the emissivity of the surface used as a reference must be known and also that of the surface of interest before quantitative data can be obtained. Table 5.1 shows typical values for a variety of materials. Complex problems can arise if the surface material transmits radiation: transmission of infrared through glass is high for wavelengths $< 3\ \mu$m and drops off rapidly with increasing wavelengths. In general this difficulty is not important in practice, and providing the thermographic equipment is calibrated at temperatures in the range of investigation, temperatures can be measured with sufficient accuracy for practical purposes. Quantification is important if decisions are to be made on investments, in improvements, in maintenance, operating procedures, improved insulation or equipment redesigns. Similarly, quantification is necessary whenever thermal imaging is used to determine the most economical time to replace components which are functioning inefficiently or dangerously.

The way in which thermography is applied to a particular problem depends upon the requirements of the investigation. Typical examples are described below. In many applications conventional photography is used in conjunction with thermography for ease and reliability of identification of the precise location of the thermal data.

5.9.1 *Thermal insulation problems*

There are many situations in which the identification of flaws in insulation forms an important part of plant maintenance and repair. Kreider and Sheahen (117) have described the use of infrared thermography for industrial heat balance calculations with particular reference to channel induction furnaces used for holding iron in a molten state prior to pouring castings: typically, heat input is electrical and heat loss mechanisms are limited to radiation and convection from the outer surface and conduction of heat through a water jacket. In the problem cited the furnace wall is 500 mm thick, comprising three courses of brick and a 30 mm thick steel shell. The function of such a furnace is to hold molten iron at about 1425°C, just above is melting temperature.

Assuming a planar model, the heat flow rate is

$$Q = \frac{\Delta T}{\dfrac{x_1}{k_2} + \dfrac{x_2}{k_2} + \dfrac{x_3}{k_3} + \dfrac{x_s}{k_s} + \dfrac{1}{H_0}}$$

where $\Delta T = T\,(\text{iron}) - T\,(\text{ambient})$, k denotes thermal conductivity, x denotes thickness, the subscripts 1, 2, 3 refer to the 3 layers of brick and s to the steel, and H_0 is the overall coefficient of heat transfer away from the surface.

Providing the thickness and thermal conductivity of each layer is known, the heat flow rate through the wall and the temperature at any point within the wall can be calculated. Thermography can be used to measure the surface temperature distribution and confirm the quantitative evaluation of heat loss. It has already been noted that the thermographic imaging system provides data which must be converted to temperature units, but once this has been accomplished it is possible to estimate the total radiation loss from the surface if the surface emissivity is known.

In practice, many of the problems are evaluated empirically. For example, in the steel industry, pouring ladles, each transporting 200 tons of liquid steel at a temperature of 1600°C, are used. Each consists of a metal shell with an insulating brick lining; perforation or cracks in the lining could be hazardous but unnecessary withdrawal from routine use would be costly. Visual inspection cannot distinguish between a serious external crack and an unimportant one. Thermography may be used to monitor the external temperature of the ladles to ensure that safe surface temperatures of about 400°C are being maintained (118).

Thermal imaging has also been used to indicate when change of refractory linings in furnaces is required. By design, a new process furnace should not lose more than 350 Btu/sq ft/hr, whereas older furnaces might lose 5 to 6 times this figure (118). Deterioration of internal refractory lining in a furnace or in a cat cracker expansion chamber can cause localized hot spots. In the latter case hot spots in excess of 1000°C compared with normal operational values of about 500°C have been measured (114).

Similarly, thermographic inspection can be used for reactors and heating columns. Although man-ways and other exposed parts might only amount to 8% of the total reactor surface, unless such parts are insulated effectively, the heat loss from these surfaces can approach the level of heat loss from the remaining 92% of the surface. Regular surveillance of these areas forms an important part of maintenance.

5.9.2 *Thermal losses from buildings and district heating systems*

The high cost of energy has emphasized the need to re-examine procedures from the construction, insulation and maintenance of buildings and facilities. The energy consumption of the average domestic dwelling in the United Kingdom (floor area 100m^2) is made up of 67% space heating, 20% water heating, 10% cooking and 3% lighting. Principal savings are therefore most likely to be made by reducing space heating losses. Heat losses from domestic dwellings have been analysed by Rogers and Waters (119). These authors compared the standards of house insulation in the United Kingdom with those used in the older Scandinavian countries. Scandinavian homes are well insulated with an average living room temperature of 21°C when temperatures are sub-zero outside. The heating system operates continuously throughout the winter months. Once a house is warm the heating requirement is minimal, and the system maintains a comfortable living temperature by just ticking over. If this fails to occur the insulation is suspect, and thermography may be used to locate the insulation defect. It is interesting that in Sweden it is not uncommon to find a bank or building society refusing to lend money on a house until it has been cleared of major insulation defects by thermography. This is best carried out in the winter months to achieve the maximum thermal gradient between inside and outside ambients (typically 15 to 20°C in the UK, and 20 to 40°C in Scandinavia). Prior to the survey the central heating is left running for 24 hours to ensure that the interior house temperature is a uniform 20°C. All windows and doors should be closed and a small pressure differential (50N/m) should be introduced between the inside and outside of the house by means of a ventilation fan. Surveys are usually carried out indoors to avoid the effects of varying climatic conditions associated with external surveillance and because of the double-skin nature of most Scandinavian dwellings. A void cavity wall insulation will result in a cold spot on the face of the interior wall approximately 2°C lower in temperature than the surrounding insulated regions. Cold spots are classified in terms of severity according to their average temperature relative to the surroundings and the area over which the cold spot extends. Each cold spot is checked subsequently for draughts. Used in this way, thermography can identify design inadequacies or faulty workmanship. It is also a useful means of establishing the validity of mathematical models employed to estimate thermal losses from structures of differing designs (120, 121).

The transfer of heat through layer structures such as those used as walls may be considered in the following way (119). If T_i is the air temperature inside the building, T_1 and T_2 are the inside and outside temperatures of the wall and T_0 is the outside air temperature then in the steady state the quantity of heat (Q) flowing through unit area of wall is described by the following equations:

$$
\begin{aligned}
Q &= H(T_i - T_1) + \sigma\varepsilon[(T_i + 273)^4 - (T_1 + 273)^4] \\
&= \frac{T_1 - T_2}{x_1/k_1} \\
&= H(T_2 - T_0) + \sigma\varepsilon[(T_2 + 273)^4 - (T_0 + 273)^4]
\end{aligned}
$$

where H is the heat transfer coefficient, σ is the Stefan-Boltzmann constant, and x and k are the thickness and thermal conductivity of the wall respectively.

The temperature is in °C and it is assumed that (ε) the emissivities of room and walls are equal.

The above equations can be approximated to:

$$Q = (H + 4\sigma\varepsilon 273^3)(T_i - T_1) = \frac{T_i - T_1}{R_1}$$

$$= \frac{T_1 - T_2}{x/k} \qquad = \frac{T_1 - T_2}{R_2}$$

$$= (H + 4\sigma\varepsilon 273^3)(T_2 - T_0) = \frac{T_2 - T_0}{R_1}$$

If $R = R_1 + R_2$ then

$$Q = \frac{T_i - T_0}{R}$$

R is called the thermal resistance, and the thermal transmittance $U = 1/R$ is the overall air to air coefficient of heat transmitted through the wall. If the structure consists of multiple layers the total thermal transmittance is made up of a series of components, each having its own resistance. The thermal efficiency of different structures may be described by comparing values of thermal transmittance. British building regulations relating to thermal insulation require specific U values in domestic buildings. The thermal transmittance for roofs (combined with ceiling) is specified to be 0.6 W/m^{2}°C and that of the walls 1.0 W/m^{-2}°C. Thermography can be used to measure U values so that compliance with building regulations can be assessed.

Whereas many of the thermal properties of building materials are known, the emissive properties are less well known. Consequently, although thermography is now an established method for investigating thermal losses, quantitative assessment depends upon the availability of values for the emissivities of the surfaces being investigated (122). American studies (123) using an aerial thermal imaging system illustrate this problem. The airborne survey was achieved with a multi-spectral scanner at an altitude of 1200 feet. Although excellent thermal images were obtained and apparent temperatures of buildings over a large urban area were recorded, environmental factors were found to influence the relation between roof temperature and insulation. These included the roofing materials, the pitch and orientation of the roof as well as external sources of radiation. Sampson and Wagner (123) developed an analytical model to interpret observed roof temperatures from flight data and found that the actual roof surface temperatures were 2 to 3°C warmer than the "apparent" temperatures recorded. Allowance must be made in such calculations for the shape of the radiating surface. For example, buildings recorded in aerial surveys which have pitched rather than flat roofs, receive

Figure 5.16 Thermogram recorded at night with an IR 18 Barr and Stroud Mark II camera showing its value for surveillance and illustrating thermal losses from buildings (Black, cold; white, hot). (Courtesy Barr and Stroud Ltd).

radiation from nearby surrounding surfaces such as adjacent buildings. In a questionnaire follow-up to one thousand homes imaged, a correlation of greater than eighty percent was obtained between predictions based on thermal data and residents reporting the existence of thermal insulation in their homes. Figure 5.16 illustrates the potential of thermal imaging for both monitoring heat loss from buildings and also its use for surveillance.

Denmark has the highest density per inhabitant of district heating systems in the world with about 16 000 km of underground hot water pipes. Thermography has proved to be an efficient tool for precise location of heat leakages from damaged or poorly insulated pipes (124). Surveys are usually carried out at night when traffic is low and the road-surface and ambient temperatures are low. By means of a specially adapted motor vehicle it is possible to examine 10–20 km of pipes per hour. A spotted fault is usually marked directly on the road or on the network map. The technique is highly cost effective because of the costs involved in visual inspection or digging up areas where a leak is suspected.

5.9.3 *Detection of malfunctioning components*

Energy loss inspections are necessary in many situations. The proper operation of steam traps and relief valves is characteristic of the sort of problem faced by maintenance departments in many industries. The main function of a steam trap is to discharge condensate while retaining steam in the system. Using thermography it is possible to examine a number of traps on the same manifold on both the discharge and inlet sides. A low temperature differential indicates if steam is blowing through or, if the inlet temperature were considerably lower, then this would suggest "backing up" of the condensate. Thermography can be used to identify problem areas and prevent unnecessary shut-downs (114, 125).

Thermography is also being used in a variety of manufacturing industries which make electrical components and appliances. The mass production of TV sets with high safety standards requires extensive testing of all circuits and mechanical and electronic components at the development stage. Thermal imaging is used for detecting overheating or malfunctioning components. The technique is useful as a means of optimizing circuit design and also it provides permanent documentation which may be used subsequently to investigate the thermal distribution resulting from component failure and its effect on the heating-up of the back cover of the TV set (116).

Infrared inspection is used extensively in the electricity generating industry and also in the maintenance of electrical equipment (124, 126). Large areas of inaccessible objects can be scanned rapidly and remotely under normal operating conditions, without danger or inconvenience. Funnel and Thelwell (115) have described its use to monitor electrical power plant and have demonstrated how it can be used to detect imminent failure of electrical components due to inadequate cooling or failure of electrical or thermal insulation. The method is particularly useful in the following investigations.

Inspection of generator stators. When large generators are constructed or overhauled it is necessary to test the integrity of the stator lamination insulation by subjecting the core to a ring flux test producing the rated voltage gradient. Imperfections in the core insulation may produce local heating. This can be detected thermographically by means of a camera mounted on the axis of the generator on a remotely controlled motorized trolley. Tests on 500 MW generators have shown that if localized hot spots of 10°C or more occur after one hour, then repair work is necessary. The laminations have to be ground away over the damaged area, or alternatively fresh insulation such as mica or epoxy resin can be forced between the laminations in the suspected area. Thermography can be used subsequently to check the efficacy of the repair.

Slipring temperature measurements. Rotors of large generators have sliprings which carry the main excitation current of several thousand amps. These sliprings are air-cooled, and if the temperatures are too high, excessive wear of carbon brushes and slipring surfaces can cause problems. The slipring surface can be monitored thermographically: temperatures over about 100°C are thought to lead to increased brush wear rates.

Transmission plant applications. The largest generators produce electricity at 25 000 volts, but for efficient transmission over long distances the voltage is increased by transformers to 132 000, 275 000 or 400 000 volts. The voltage is reduced again by subsidiary transformers for distribution to consumers. To meet these requirements, transmission plant must include thousands of miles of grid lines, many substations with numerous large transformers, busbars and circuit breakers all of which are at high voltage. Thermal surveying has been used since 1965 and now forms an important part of the United Kingdom's Central Electricity Generating Board's maintenance programme. The technique is used for three purposes: firstly, for testing of overhead transmission lines. Aerial thermography is used to locate hot joints and fittings on overhead transmission lines. There are about 14 000 km of 275 kV and 400 kV lines in service with the CEGB, with about twenty joints per kilometre. The conductors consist of a steel core wrapped with aluminium and there are joints between drum lengths and at tension towers. The joints can deteriorate in service due to defective components or corrosion. Helicopter patrols about 80 m from the line and slightly above the earth wire height have proved to be a very valuable tool in preventive maintenance. It is possible to survey 300 km of circuit a day in search of faults which show as hot spots when the line is loaded. Initial tests with simulated faults showed that temperature rises of 3°C can be identified under normal flying conditions, Maintenance patrols are flown annually from October to April when the load on the grid is greatest, and repair work is performed in the summer months. Typically, faults causing localized hot spots with temperatures greater than 30°C are repaired immediately, whereas those with temperatures 5–30°C are repaired on the first possible occasion. Hot spots less than 5°C are dealt with at the next planned overhaul (127).

Secondly, similar ground-based surveys are undertaken at substations which transform supplies to lower voltages for redistribution. In these substations are many connections and joints which link the overhead lines, cables, measurement and protection equipment and transformers via busbars. Inspections are made regularly with the aid of an infrared camera system and a special purpose vehicle. Finally, faults in high voltage insulating material often produce discharges which can be recorded by special detection equipment, but the precise location of the faults is difficult. Thermal imaging has been used to show non-uniform temperature distributions when faults are present, and this data is helpful for identifying their position. Typical failure of a 33kV surge divertor is illustrated in Plate I, D: the thermogram shows a 4.4°C increase above normal.

5.9.4 *Thermography as a design aid*

To be of value, investigations such as thermal stress analysis must be made under various conditions with different loads; thermal imaging is ideally suited for this purpose. The technique has been applied in wind-tunnel tests of aircraft (128), the design and manufacture of aircraft windshields (129) and safety glass windows for the automobile industry (116). In the latter case the method may be used to monitor glass production and to ensure that the design of heated rear windows of cars is adequate to distribute heat uniformly over the window (116). In Scandinavia, SAAB car manufacturers have designed a driver's seat which is heated automatically. When the temperature of the seat falls below + 14°C, integral heating coils are activated when the car ignition key is switched on (130). The resultant effect may be monitored thermographically to ensure compliance with design specification and customer requirements.

Clothing design is another area of interest, both from a commercial standpoint and physiologically. The efficacy and efficiency of winter and summer clothing may be assessed thermographically in terms of heat loss and factors such as fashion, fabric thickness and manufacturing costs can be analysed in relation to this important physical parameter. Specialist type clothing (garments made from microfibres, high performance insulating materials, those made from materials with highly reflective surfaces and neoprene/nylon diving suits) as well as common clothing for business or sport have been the subjects of thermographic scrutiny (130). Thermal investigations have also been made into the efficacy of bed quilts. In a constant ambient temperature of 18°C, using a thermal mannequin as a heat source and quilts made with 1 kg of uniform quality down, temperature difference between the coldest and warmest areas of well-designed quilts was found to be 1.5°C, whereas in some quilts this difference can be as high as 3.5°C (130).

5.10 Military applications

The history of infrared thermography is linked closely with the development of detectors for military purposes. Medical and industrial applications of the technique resulted directly from the use of these detectors in specially constructed scanning systems. Because of the advantages of high resolution, passive detection and observation of targets in darkness and poor visibility,

Figure 5.17 Thermogram of tank taken at night with an IR 18 Barr and Stroud Mark II camera (Courtesy Barr and Stroud Ltd).

thermal imaging systems are still important militarily. Such systems are superior in performance to image intensifiers for detecting targets at long range or in adverse weather. The high thermal and spatial resolutions of relatively small, real-time imaging systems allow thermal images to be used in many ways, including anti-tank missile guidance applications, tank-mounted and other vehicle-mounted observation and aiming systems, the acquisition and tracking of low-flying aircraft, and in high-performance forward-looking systems for aircraft and helicopters. Specific requirements for such applications include compactness of design, robustness of construction, high resolution, and very fast scanning speeds. Perception of camouflaged personnel, buildings or military equipment and the imaging of terrestrial objects by aerial and satellite surveillance are also possible (7).

Because thermal imaging can be carried out remotely and at high speed, the technique has been used in the development and assessment of arms and equipment. The measurement of gun-barrel temperatures after test shooting at various firing rates is typical of such an application (131).

The use of infrared weapon sights is not dependent upon clear visibility and such systems may be used in total darkness or dense smoke. Passive reconnaissance and detection of objects in the air, on sea or land are of great practical importance. Figure 5.17 shows a thermogram of a tank recorded at night with an IR 18 Barr and Stroud Mark II system, equipped with a Mullard Sprite detector. Thermal "fingerprints" can supplement other

Figure 5.18 Thermogram of two sea vessels recorded at 4 km with an IR 18 Barr and Stroud Mark II camera (Courtesy Barr and Stroud Ltd).

surveillance data: for example, vessels at sea in poor visibility may be located thermographically and their thermal characteristics can reveal the type of ship and whether or not the vessel is under steam; in figure 5.18 the vessel is significantly warmer midships. The study of radiant emission from jet aircraft plumes has led to a better understanding of missile guidance problems and to improvements in infrared seeking heads (130, 131).

5.11 Microwave thermography systems

We have noted already that the emission of thermal radiation by a body is described by Planck's law and that at 30°C peak emission occurs at about 10 μm. Figure 5.19 shows that the intensity of this radiation is about 10^8 times greater than that of 10 cm wavelength radiation. At microwave frequencies where $h\nu \ll kT$ the exponential in Planck's formula may be expanded to give the Rayleigh-Jeans expression for the intensity E of this radiation:

$$E(\nu, T) = \varepsilon(\nu)\frac{2\pi kT\nu^2}{c^2}$$

where $\varepsilon(\nu)$ is the emissivity which depends on the dielectric properties of the emitting and receiving media, k is Boltzmann's constant, c is the speed of

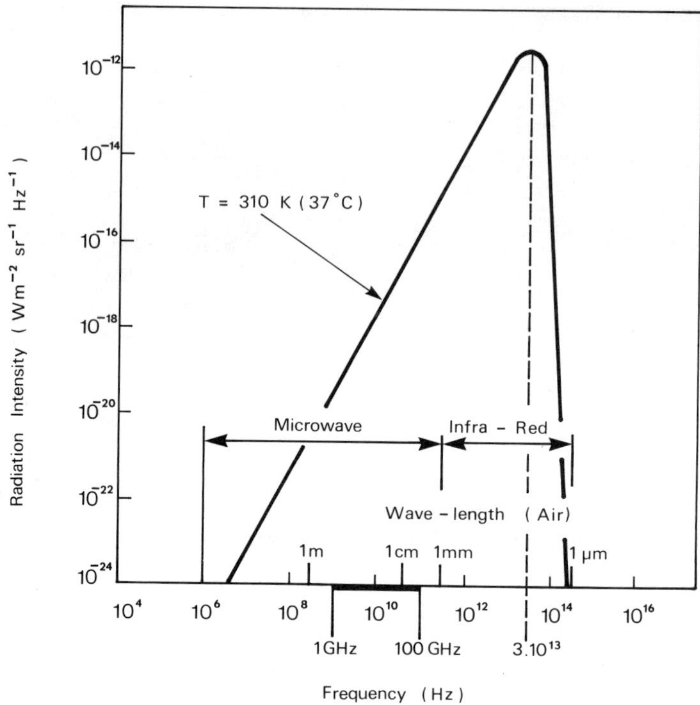

Figure 5.19 The variation of radiant emission from a human body at 37°C with frequency.

light, T is the temperature of the object and ν is the frequency of the emitted radiation. Thus for microwave emission, the radiant power passing through unit area into a unit solid angle in a unit of interval of frequency $d\nu$ may be considered to be proportional to the temperature of the emitter. Microwave thermography (sometimes called "centimetre and millimetre thermography") can be used to determine subcutaneous temperatures and locate pathological conditions associated with temperature change (132, 133, 134). The method is used conjointly with infrared thermography which provides surface data only; microwave thermography is largely dependent upon the permittivity of the radiating tissues and the properties of the tissues through which the radiation passes, and provides additional sub-surface data. The values of the permittivity of skin and muscle at cm and mm wavelengths, as well as radiometric measurements of the skin-to-air emission, show that the emissivity increases from a relatively low value of 0.5 at 3 GHz to 0.96 at 30 GHz (135). The radiation penetration within the body depends strongly on the absorption properties of high- and low-content water tissue. At 3 GHz the penetration depth in fat is about 5 cm and about 0.8 cm in muscle or skin. The corresponding wavelengths are 10 cm in free space and about 5 cm and 1.6 cm in fat and muscle respectively. The penetration depth decreases with increasing frequency and at frequencies above 10 GHz escaping radiation originates in a skin layer < 1 mm deep (136). This implies that imaging at these frequencies should give results similar to those of infrared thermography.

Although microwave emissions from the human body at 37°C are very weak, microwave radiometers (originally developed for radio-astronomy) are capable of detecting this low-intensity radiation and are sufficiently sensitive to measure power changes equivalent to temperature differences of 0.1°C. Penetration depths in tissue for radiation with a frequency < 10 GHz are significantly greater than those for infrared radiation. However, radiation collected by a radiometer on the body surface has been emitted by tissues at various depths and analysis of the emission pattern is a complex problem (132, 137).

Two types of radiometer have been developed for medical investigations: contact antennas (136) and remote sensing radiometers with focusing reflectors (135). Contact antennas are used for detecting 1–3 GHz radiation. The receiver probes have apertures about 1 × 2 cm. These are filled with a dielectric material of appropriate permittivity to accomplish matching, and are placed in contact with the skin surface. The equipment is less complex than that used for scanning investigations and is not suitable for imaging techniques. In addition, the technique suffers from the following limitations.

1. Whereas the spatial resolution on the skin surface can be relatively good (of the order of a cm or so), the resolution at a depth deteriorates rapidly and precludes the identification of small localized areas of high temperature.
2. The incremental signal received by an antenna resulting from a tumour (or hot spot) at a depth is relatively small compared with radiation received from other tissues close to the skin surface and also warmer tissues at greater depth.

These problems are accentuated further by large signal changes caused by varying match at different skin locations as the probe is moved over the skin surface. The natural variations of peripheral tissue composition within the range of an antenna between different regions of a body or between different individuals results in a spread of the wave impedance at the body's surface with a corresponding variation in emissivity. Reduction of errors arising from this effect can be achieved if the temperature of the receiver and object are made equal. Luedcke (138) has described a method of achieving this effect with a receiver which measures the difference between the incoming and outgoing radiation flux. An integrating servo amplifier is used to control a variable noise source to add excess thermal radiation to the receiver outgoing radiation flux. A self-balancing radiometer of this type is useful for eliminating errors due to mismatch.

In spite of its limitations, contact microwave temperature studies have been used in conjunction with infrared thermography in the study of breast cancer. Myers *et al.* (136) have used Dicke-switched type radiometers which operate at 1.3, 3.3 and 6.0 GHz. These authors have used 1.3 and 3.3 GHz and infrared thermography to examine large numbers of women with breast disease. Their work suggests that there is some advantage in using infrared and microwave thermography in concert.

The second type of radiometry employs remote sensing by focused apertures like lenses or reflectors (134, 135). This method is better suited to higher frequencies because of the design requirements for an aperture diameter of many wavelengths. Typically 30 GHz and 68 GHz detector systems have been used: the emitted radiation is focused by a reflector into a horn which contains the r.f. portion of a sensitive Dicke radiometer. Being a non-contact system it allows scanning of the surface to be employed. Unfortunately, going to these higher frequencies decreases penetration depths and at frequencies about 10 GHz, radiation absorption within the body is such that escaping thermal emission must originate in the most superficial tissues. This disadvantage is compensated by improvements in resolution and power factors. At 9 GHz, Edrich (135) has predicted that a 1.5°C, 2 cm deep hot spot of 2 cm diameter would cause an increase in antenna temperature of about 0.1°C. Elevations of this magnitude can be detected by remotely sensing radiometers combined with scanning and processing systems to provide thermal images.

The technique has been used clinically for imaging breast cancer (134), thyroid nodules (139), and brain lesions (133). It has also been used to monitor inflammatory conditions such as arthritis. This type of thermography provides different information about subcutaneous temperatures to that of infrared techniques. It is limited severely by lack of spatial resolution and factors such as tissue composition and spatial distribution, as well as the dielectric properties of the body, all of which influence the thermal record and should be known before the measured temperature pattern can be interpreted accurately. These fundamental limitations are important when the technique is used for diagnostic imaging, but are less so when microwave thermography is used to monitor induced hyperthermia. In this case, localized tissue temperatures are increased to about 43°C; the heating might be localized and superficial or deep within the body. To avoid implanting thermocouple probes subcutaneously, microwave techniques are being used to monitor temperatures and control heating procedures (140).

Microwave systems that rely on natural thermal radiation from a scene are also used for mapping the earth and moon. Compared with lower frequencies the microwave band (1 GHz—1 THz corresponding to wavelengths from 300 mm to 300 μm) enables useful angular resolutions to be obtained using reasonably sized apertures, while it offers extremely low atmospheric attenuations compared with higher frequencies. A typical aerial scan from a low-speed aircraft flying at medium altitude will yield a minimum temperature differential sensitivity of about 2°C while a representative lunar mapping radiometer will have a sensitivity of some 10°C per second of integration (141).

References

1. R. A. Smith, *The Physical Principles of Thermodynamics*, Chapman and Hall (1952).
2. R. A. Smith, F. E. Jones and R. P. Chasmar, *The Detection and Measurement of Infrared Radiation*, eds. W. Jackson, H. Frohlich, N. Mott, Oxford University Press (1968).
3. J. A. Jamieson, R. H. McFee, G. N. Plass, R. H. Grube, and R. G. Richards, *Infrared Physics and Engineering*, McGraw-Hill, (1963).
4. J. Steketee, *Phys. Med. Biol.* **21** (1976) 920–930.
5. P. W. Kruse, L. D. McGlauchlin, R. B. MacQuistan, *Elements of Infrared Technology*, Wiley (1968).
6. *Aga Thermovision Model* 680 *Medical Operating Manual*, Aga Infrared Systems AB (1975).
7. W. L. Wolfe, (ed.), *Handbook of Military Infrared Technology*, Washington DC: Naval Research Department of the Navy, 1965, updated edition: *The Infrared Handbook*, ed. W. L. Wolfe and O. O. Zissis (1978).
8. J. Steketee, *Phys. Med. Biol.* **18** (1973) 686.
9. D. J. Watmough, P. W. Fowler, R. Oliver, *Phys. Med. Biol.* **15** (1970).
10. C. J. Martin, and D. J. Watmough, *Acta Thermographica* **2** (1977) 18–22.
11. J. A. Clark, *Acta Thermographica* **1** (1976) 138–141.
12. K. Lloyd-Williams, F. J. Lloyd-Williams, and R. S. Handley, *Lancet* **2** (1961) 1378.
13. G. W. McDaniel, and D. Z. Robinson, *Appl. Opt.* **1** (1962) 311.
14. R. B. Barnes, in *Ann. N.Y. Acad. Sci.* **121** (1964) 34–48.
15. W. D. Lawson, S. Neilsen, E. H. Putley, and A. S. Young, *J. Phys. Chem. Solids* **9** (1959) 325.
16. R. W. Astheimer, and F. Schwarz, *Appl. Opt.* **7** (1968) 1687–1695.
17. R. Watton, *Ferroelectrics* **10** (1976) 91–98.
18. D. E. Burgess, in *Recent Advances in Medical Thermography*, Proc. 3rd E. A. T. Congress, Bath, U.K. (1982), Plenum, in press.
19. R. Watton, D. Burgess, and P. Nelson, *Infrared Physics* **19** (1979) 683–688.
20. P. Newman, M. Davison, and W. B. James, *Acta Thermographica* **4** (1979) 132–136.
21. W. L. Wolfe, in *Thermography and its Clinical Applications*, ed. H. E. Whipple, *Ann. N.Y. Acad. Sci.*, **121** (1964) 69.
22. C. H. Jones, in *Scientific Basis of Medical Imaging*, ed. P. N. T. Wells, Churchill Livingstone (1982).
23. W. D. Lawson, in *Electronic Imaging*, T. P. McLean and P. Schagen, Academic Press (1979) 325–364.
24. J. L. Fergason, *Appl. Opt.* **7** (1968) 1729–1737.
25. T. W. Davison, K. L. Ewing, J. Fergason, M. Chapman, A. Can, and C. C. Voorhis, *Cancer* **29** (1972) 1123.
26. W. E. Woodmansee, *Appl. Opt.* **7** (1968) 1721–1727.
27. S. U. Flesch, in *Recent Advances in Medical Thermography*, Proc. 3rd E. A. T. Congress, Bath, U.K. (1982), Plenum, in press.
28. Ch. Gros, M. Gautherie, P. Bourjat, and F. Archer, *Ann. Radiol.* **13** (1970) 333.
29. W. Gordenne, *J. Belge Radiol.* **60** (1977) 139.
30. J. Tricoire, L. Mariel, and J. P. Amiel, *J. Radiol. Electrol.* **53** (1972) 13.

31. R. Pochaczevsky and P. H. Meyers *Acta Thermographica* **4** (1979) 8–16.
32. G. W. Gray and D. G. McDonnell *Mol. Cryst. Liq. Cryst.* **48** (1978) 37.
33. D. G. McDonnell and I. Sage, in *Recent Advances in Medical Thermography*, Proc. 3rd E. A. T. Congress, Bath, U.K. (1982), Plenum, in press.
34. K. G. Archer in *Recent Advances in Medical Thermography*, Proc. 2nd E. A. T. Congress. Bath, U.K. (1982), Plenum, in press.
35. M. Tonegutti, L. Acciarri, and A. Racanelli, *Acta Thermographica*, Suppl. 3 (1980).
36. R. Pochaczevsky, C. E. Wexler, P. H. Meyers, J. A. Epstein and J. A. Marc *J. Neurosurg.* **56** (1982) 386–395.
37. F. Lelik and G. Kezy, *Acta Thermographica* **4** (1979) 24–29.
38. F. Lelik, G. Kézy and D. Solymossy, *Acta Thermographica* **2** (1977) 13–17.
39. K. Atsumi (ed.), *Medical Thermography*, University of Tokyo Press (1973).
40. E. H. Putley, in *Recent Advances in Medical Thermography*, Proc. 3rd E. A. T. Congress, Bath, U.K. (1982), Plenum, in press.
41. Mullard Technical Publication M82–0099, *Electronic Components and Applications* **4** (4) 1982.
42. C. T. Elliott, *Electron. Lett.* **17** (1981) 312.
43. Y. Matsui, M. Okuyama, N. Fujita, and Y. Hamakawa, *J. Appl. Phys.* **52** (1981) 5107.
44. R. P. Clark, in *Recent Advances in Medical Thermography*, Proc. 3rd E. A. T. Congress, Bath, U.K. (1982), Plenum, in press.
45. T. J. Love in Ann. N.Y. Acad. Sci. Conference on thermal characteristics of tumours: applications in detection and treatment, New York, 1979, *Ann. N.Y. Acad. Sci.* **335** (1980) 429–437.
46. R. K. Jain, in Ann, N.Y. Acad. Sci. Conference on thermal characteristics of tumours: applications in detection and treatment, New York, 1979, *Ann. N.Y. Acad. Sci.* **335** (1980) 48–64.
47. H. H. Pennes, *J. Appl. Physiol.* **1** (1948) 93–122.
48. J. D. Haberman, J. E. Francis, and T. J. Love, *Radiology* **102** (1972) 341–348.
49. L. M. Torrell, and S. K. Nilsson, *Phys. Med. Biol.* **25** (1980) 85.
50. C. H. Jones, and J. W. Draper, *Br. J. Radiol.* **43** (1970) 507.
51. J. R. Hill, and K. A. Rahimtulla, *J. Physiol.* **180** (1965) 239.
52. K. J. Collins, in *Recent Advances in Medical Thermography*, Proc. 3rd E. A. T. Congress, Bath, U.K. (1982), Plenum, in press.
53. E. Rylander,(1972). *Acta Paediat. Scand.* **61** (1972) 1.
54. N. J. Rothwell, and M. J. Stock, *Nature* **281** (1979) 31–3.
55. M. Gautherie, D. Gros, and Ch. Gros, *Acta Thermographica* **2** (1977) 23–37.
56. J. Steketee, *Acta Thermographica* **4** (1) (1979) 43–47.
57. H. J. Isard, W. Becker, and R. Shilo, *Am. J. Roentgen*, **115** (1972) 811.
58. A. M. Stark, and S. Way, *Cancer*, **33** (1974) 1671–1679.
59. M. Gautherie, E. Grosshans, and J. Juillard, *Supplement Medicamundi*, (1979).
60. N. T. Johansson, N. Bjurstam, K. Hedberg, A. Hultborn, and C. Johnsen, *Acta Chirurgica Scandinavica*, Supplementum 460 (1976).
61. C. H. Jones, and J. B. Davey, in *Medical Oncology*, (ed.) K. D. Bagshawe Blackwell Scientific (1975) 215.
62. E. Comhaire, R. Monteyne, and M. Kunnen, *Fertil. and Steril.* **27** (1976) 694–6.
63. R. H. Gold, R. M. Ehrlich, B. Samuels, A. Dowdy, and R. T. Young, *Radiology* **122** (1977) 129.
64. E. F. J. Ring *Acta Thermographica* **1** (1976) 67.
65. M. Gautherie, P. Haehnel, and Ch. Gros, *Biomedicine* **22** (1975) 416–427.
66. C. J. Martin, *Acta Thermographica* **4** (1979) 69–7.
67. C. Gros, M. Gautherie, and P. Bourjat, in *Thermography* (Proc. 1st Europ Congr. Amsterdam 1974) *Bibl. Radiol.* **6**, Karger (1975) 77–90.
68. K. D. Patil, J. R. Williams, and K. Lloyd-Williams, *Br. Med. J.* **1**, (1970) 195.
69. E. D. Cooke, and M. F. Pilcher, *Br. J. Surg.* **61**, (1974) 971.
70. N. Hendler, S. Uematesu and D. Long, *Psychosomatics* **23**, 3 (1982) 283–287.
71. M. E. J. Hackett, *Br. J. Plastic Surgery* **27** (1974) 311–317.

72. G. Buwalada, in *Medical Thermography* (Proc. Boerhaave Course for Postgraduate Medical Education, Leiden, 1968) *Bibl. Radiol.* **5**, Karger (1969) 178.
73. G. Buwalada, in *Medical Thermography* (Proc. Boerhaave Course for Postgraduate Medical Education, Leiden, 1968) *Bibl. Radiol.* **5**, Karger (1969) 122.
74. H. E. Whipple (ed.), *Thermography and its Clinical Applications, Ann. N.Y. Acad. Sci.* **121** (1964).
75. S. F. C. Heerma van Voss, and P. Thomas (ed.), *Medical Thermography* (Proc. Boerhaave Course for Postgraduate Medical Education, Leiden, 1968) Karger (1969).
76. N. J. M. Aarts, M. Gautherie, and E. F. J. Ring (eds.) *Thermography* (Proc. 1st Europ. Cong. on Thermography, Amsterdam, 1974) Karger (1975).
77. R. E. Woodrough, *Medical Infra-red Thermography*, Cambridge University Press (1982).
78. M. Gautherie, and E. Albert, (eds.) *Biomedical Thermology* (*Progress in Clinical and Biological Research* **107**) R. Alan Liss Inc., (1982).
79. *Recent Advances in Medical Thermography* (Proc. 3rd E. A. T. Congress, Bath, U.K. 1982) Plenum, in press (1983).
80. V. A. Spence, W. F. Walker, I. M. Troup, and G. Murdoch, *Angiology* **32** (3) (1981) 155–169.
81. E. D. Cooke, Supplement 1 to *Acta Thermographica* (1978).
82. W. G M. Ritchie, M. S. Lapayowker, and R. L. Soulen, *Radiology* **132** (1979) 321–329.
83. C. H. Jones, and W. F. Hendry *Acta Thermographica* **4** (1) (1979) 38–43.
84. B. D. Fornage, J. L. Valeyne, P. L. Lemaire, and B. M. Lardennois, *Acta Thermographica* **5** (1) (1980) 38.
85. J. W. Draper and C. H. Jones, *Br. J. Radiol.* **42** (1969) 401.
86. C. H. Jones, W. P. Greening, J. B. Davey, J. A. McKinna and V. J. Greeves, *Br. J. Radiol.* **48** (1975) 532.
87. R. N. Lawson, *Can. Med. Ass. J.* **75** (1956) 369.
88. M. Gautherie, in *Ann. N.Y. Acad. Sci.* **335** (1980) 383.
89. C. H. Jones, in *Thermography* (Proc. 1st Europ. Cong., Amsterdam, 1974) *Bibl. Radiol.* **6**, Karger (1975) 57.
90. M. S. Lapayowker, I. Barash, R. Byrne, C. H. J. Chang, G. Dodd, C. Farrell, J. D. Habermann, H. J. Isard and B. Threatt, *Cancer* **38** (1976) 1931.
91. M. Gautherie, *Cancer* **45** (1980) 51.
92. R. Amalric, J. M. Spitalier, D. Giraud, and C. Altschuler, in *Thermography*, (Proc. 1st Eur. Congr. Amsterdam, 1974) *Bibl. Radiol.* **6**, Karger (1975).
93. M. Gautherie, in *Biomedical Thermology*, (*Progress in Clinical and Biological Research* **107**), eds. M. Gautherie and R. Albert, R. Alan Liss Inc., (1982).
94. A. J. Collins, E. F. J. Ring, J. A. Cosh, and P. A. Bacon, *Ann. Rheum. Dis.* **33** (1974) 113.
95. E. F. J. Ring, in *Thermography*, (Proc. 1st Europ. Congr. Amsterdam, 1974) *Bibl. Radiol.* **6**, Karger (1975) 97.
96. R. Mladick, N. Georgiade, and F. Thorne, *Plastic Reconst. Surg.* **38** (5) (1966) 12.
97. H. T. Bergholdt, and P. W. Brand, in *Proc. of the Ann. Meeting of the American Thermographic Soc.* (1973) 211.
98. L. Acciarri, L. Colognese, A. Pizzoli, C. Boscaro, and R. Maso, *Acta Thermographica* **2** (3) (1977) 166.
99. C. E. Wexler, *J. Neurol. Orthop. Surg.* 1 (1979) 37–41.
100. R. Pochaczevsky, F. Feldman, *A. J. N. R.* **3** (1982) 243–250.
101. P. Carnochan, and C. H. Jones, in *Recent Advances in Medical Thermography* (Proc. 3rd E. A. T. Congress, Bath, U.K. 1982) Plenum, in press (1983).
102. K. Cena, and J. A. Clark, *Phys. Med. Biol.* **18** (1973) 432.
103. K. Cena, and J. A. Clark, *Acta Thermographica* **3** (3) (1978) 135–141.
104. B. Strömberg, in *Thermography* (Proc. 1st Europ. Congress. Amsterdam, 1974) *Bibl. Radiol.* **6**, Karger (1975) 231–236.
105. L. W. Nichol, and J. Lamar, *Appl. Opt.* **7** (9) (1968) 1757–1762.
106. *Proc. of the Tenth Int. Sym. on Remote Sensing of the Environment*, University of Michigan (1975).
107. C. C. Goillot, in *Thermography* (Proc. 1st Europ. Congr., Amsterdam, 1974) *Bibl. Radiol.* **6**, Karger (1975) 237.

108. *Int. J. Remote Sensing* **3** (3) (1982) 350.
109. G. M. Lechi, *Acta Thermographica* **2** (1) (1977) 4–12.
110. R. Cassinis, C. M. Marino, and A. M. Tonelli, in U.N. Symposium on the Development and Utilization of Geothermal Resources, Pisa, *Geothermics* **2** (1) Special Issue 2 (1970).
111. G. A. Pieters, in *Thermography* (Proc. Ist Europ. Congr., Amsterdam, 1974) *Bibl. Radiol.* **6**, Karger (1975) 210.
112. C. W. Hurley, and K. G. Kreider, *Applications of Thermography for Energy Conservation in Industry*, NBS Technical Note 923, U.S. Department of Commerce (1976).
113. L. Green, Report: National Conference on Condition Monitoring (1977).
114. *The Infrared Observer* **5** (1980).
115. I. R. Funnell, and M. J. Thelwell, *Phys. Technol.* **9** 141–147 (1978) 141.
116. *The Infrared Observer* **6** (1981).
117. K. G. Kreider, and T. P. Sheahen, *Use of Infrared Thermography for Industrial Heat Balance Calculations*. NBS Technical Note 1129, U.S. Department of Commerce (1980).
118. *The Infrared Observer* **5** (1981).
119. L. M. Rogers, and M. S. Waters, *Heat Losses from Domestic Dwellings*, Eurotest Technical Bulletin E25 (1975).
120. R. A. Grot, and D. T. Harrje, IRIE Proceedings (1976) 103.
121. P. A. D. Mill, SPIE 254, *Thermosense* **3** (1981) 121.
122. M. S. Waters, and L. M. Rogers, *Thermography—an aid to energy conservation in the process industry*. Eurotest Technical Bulletin E26 (1976).
123. R. E. Sampson, and T. W. Wagner, *IRIE Proc.* (1976) 89–97.
124. *The Infrared Observer* **9** (1982).
125. *The Infrared Observer* **3** (1980).
126. E. J. Galto, *Plant Engineering* (1977).
127. *The Infrared Observer* **1** (1979).
128. H. Thomann, and B. Frisk, *Int. J. Heat Transfer* **11** (1968) 819–826.
129. *The Infrared Observer* **7** (1981)
130. *The Infrared Observer* **2** (1980).
131. *Military Applications of Thermovision*, Aga Publication (1970).
132. Y. Leroy, in *Biomedical Thermology* (*Progress in Clinical and Biological Research* **107**), (eds. M. Gautherie, E. Albert), R. Alan Liss Inc. (1982) 485.
133. P. Thouvenot, J. Robert, A. Mamouni, and C. Renard, in *Biomedical Thermology* (eds. M. Gautherie and E. Albert), R. Alan Liss Inc. (1982) 501.
134. J. Edrich, W. E. Jobe, R. K. Cack, W. R. Hendee, C. J. Smyth, M. Gautherie, Ch. Gros, R. Zimmer, J. Robert, P. Thouvenot, J. P. Escanye, and C. Itty, in *Ann. N. Y. Acad. Sci.* **335** (1980) 456.
135. J. Edrich, *J. Microwave Power* **14** (1979) 95.
136. P. C. Myers, N. L. Sadowsky, and A. H. Barrett, *J. Microwave Power* **14** (1979) 105.
137. S. Caorsi, in *Recent Advances in Medical Thermography*. (Proc. 3rd E. A. T. Congress, Bath, U.K. 1982) Plenum, in press (1983).
138. K. M. Luedcke, J. Koehler, and J. Kanzenbach, *J. Microwave Power* **14** (1979) 117.
139. J. Robert, J. Edrich, P. Thouvenot, M. Gautherie, and J. M. Escanye, *J. Microwave Power* **14** (1979) 131.
140. D. D. N'Guyen, A. Mamouni, Y. Leroy and E. Constant, *J. Microwave Power* **14** (1979) 135.
141. R. Voles, in *Electronic Imaging*, (eds.) T. P. Mclean, and P. Schagen, Academic Press, (1979) 395–418.

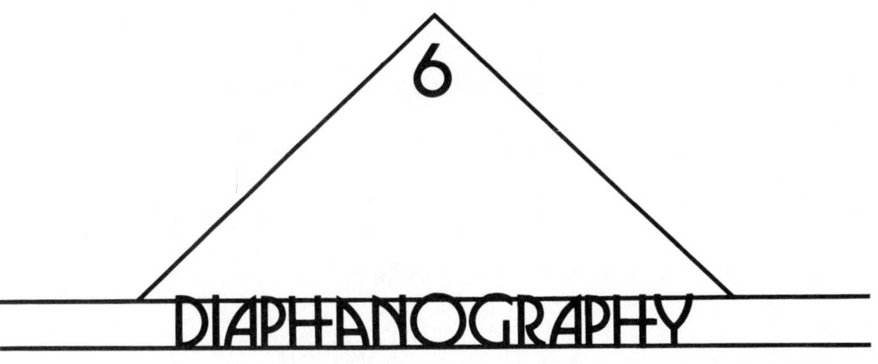

D.J. Watmough

6.1 Introduction

It has long been recognized (1) that skin and subcutaneous tissues are partially transparent in the visible and near infrared region of the spectrum. The percentage transmission varies with wavelength, with tissue type, with pigmentation and with thickness. Culter (2) utilized visible light to investigate lesions in the female breast. His pioneering work was not substantially developed until the 1970s with the work of Gros *et al.* (3) and Ohlsson *et al.* (4), although Bailey (5) had recognized and recommended trans-illumination as the test par excellence for cysts in the breast. Transillumination is also valuable for investigations of the brain in newborn infants. The scrotum (6) is another site where transillumination can be utilized to distinguish between solid and cystic lesions. Fluid-filled lesions, provided that the fluid is not blood, tend to transmit more light than the surrounding tissues and therefore, in a darkened room, the cyst is indicated by a region of increased brightness.

In the case of breast investigations, the advent of X-ray mammography with its ability to reveal micro-calcifications, was a major factor in turning attention away from transillumination. However, more recently the re-cognition (7) that diaphanography could be a valuable supplement to mammography has caused renewed interest. In addition the concern (8), (9) over the small but cumulative radiation dose to the breast in screening clinics where mammography is utilized repeatedly has led to a search for alternative tests which are based on non-ionizing forms of radiation. Strax (10) points out that diaphanography is a possible candidate and notes that attempts are being made to bring about technical improvements aimed at increasing the spatial resolution to the point where preclinical non-palpable lesions may be recognized.

6.2 Equipment for transillumination

A commercial company, Sinus Medical Equipment AB of Sweden, about 1977, began marketing their equipment, which was specifically designed to image and record breast lesions. It must be said that the Sinus Company have done a great deal to promote knowledge of and interest in the technique both in Europe and in America. The latest equipment tested by the author comprised a power supply, torch with a curved light pipe and a 35 mm camera supported on a movable wall mounted arm. The torch contained a light bulb, a xenon flash tube, a fan and heat-absorbing glass. The patient, sitting on a seat capable of rotation, leans slightly forward and the torch is applied to the underside of the pendulous female breast. The clinician carrying out the examination sits in front of the patient in a darkened or preferably blacked-out room. Light from the bulb at an intensity chosen to match the tissue thickness is used for a visual inspection. Areas of increased or decreased brightness are looked for. Dark areas are taken to imply the probability of a malignant lesion although blood-filled cysts, sites of aspiration and also bruises can give a similar appearance. Bright areas are taken to indicate the presence of fluid-filled cysts. Plate II, A and B, were made using equipment supplied by Sinus Medical Equipment AB of Sweden. Plate II, A, shows an opacity situated adjacent to the nipple, and B shows the typical area of increased brightness associated with a cyst. Whenever there is an indication of a lesion by transillumination, a further investigation by X-ray mammography and biopsy should be carried out. Film records are made with 35 mm infrared colour slide film as recommended by Ohlsson *et al.* (4). The camera is focused on the superior surface of the breast by an optically projected image generated within the camera. A footswitch actuates the xenon flash tube in the torch and also the synchronized shutter of the camera. Three images (under-, normal and over-exposed) are recommended for each view of the breast to take care of variations in tissue density and thickness. Patient identification is also recorded on the film and coded by a numeric keyboard incorporated in the main console.

Hussey *et al.* (5), while emphasizing the value of diaphanography when used as a supplement to mammography in a diagnostic situation, also pointed out a major limitation of the equipment. The photographic processing of infrared colour films takes about 1 hour but it is sometimes several days before the whole film is exposed, because of fluctuations in the number of patients attending for investigation. Some centres send the film away for central processing. Photographic records tend not to be available for at least a week after exposure, by which time clinical decisions may have had to be made. Unsatisfactory images cannot always conveniently be repeated after the patient has left the clinic. Thus the test is less valuable than might otherwise be the case. Ohlsson claims that infrared colour film is necessary to provide differential diagnostic information, although as far as the author is aware this has not been confirmed by other authors. Di Maggio (11) used a different apparatus and fast conventional colour film and claims to detect opacities in over 90% of malignant lesions, but makes no comment on his ability to achieve differential diagnosis. Out of 39 patients with carcinoma, Hussey *et al.* (7) using the Sinus equipment, found 33 true positives by mammography and 30 by diaphanography. The false negatives were 6 and 9 respectively. In 90 simple lesions, mammography correctly diagnosed 80 and diaphanography 58. The false negative rate was 10 and 32 respectively. The authors started by employing infrared thermography in addition, but found it much less satisfactory than diaphanography so that its use was discontinued.

6.3 Physical basis of diaphanography

6.3.1 *Optical properties of tissues*

There is a paucity of reliable information about the optical properties of biological tissues, but what is available is sufficient for the design of a torch suitable for demonstrating breast lesions. Cartwright (1) demonstrated that at 860 nm, cheek 5 mm thick transmitted 14% of the incident light. Peak transmission of 20% was at 1150 nm. Cartwright commented on the similar shape of the transmission curve for bacon fat. The objective of the work was to define the wavelength range for optimal heating of living tissues. Beyond 1400 nm, water within the tissues is very strongly absorbing. Hardy and Muschenheim (12) confirmed that peak transmission through a 1 mm thick specimen of skin and subcutaneous tissue occurred at about 1100 nm but the magnitude was reported to be 1%. Subsequently Hardy *et al.* (13) reported new measurements in which the percentage transmission was 60% at 0.43 mm thickness and 15% at 1.6 mm. The authors concluded that in the region 800 nm to 2000 nm, skin is almost non-absorbing, except for the absorption due to water and scattering effects. The output of the incandescent filament-type bulb is in the visible and near infrared region of the spectrum. Its output can be tailored by suitable choice of filters to match the transmission characteristics of tissues. A cut-off filter to absorb radiation beyond 1000 nm reduces the possibility of burning the skin.

The low values of percentage transmission through thick specimens of tissue explain the need for a xenon flash tube, with high transient light output, to facilitate exposure times of 1/60 second as used in the Sinus equipment. A short exposure time is desirable to avoid blurring due to patient movement.

Colins and Harper (14) have investigated the transmission properties of various grades of Opal perspex (ICI Welwyn Garden City, Herts, England). This material diffuses white light and can be chosen according to grade and thickness to have a well-defined percentage transmission. Two parallel sheets 'sandwiching' opaque spheres or strips can be used as a simplified model of lesions in breast tissue (15). The torch, placed against one sheet, projects a shadow on to the front opal surface. The model permits investigation of the effect of exit pupil diameter, light intensity and other details of the torch construction.

6.3.2 *Diaphanography: mechanism responsible for the images*

An investigation (16) of the transmission properties of slabs of material taken at operation from lesions in the female breast failed to reveal a clear distinction between transmission characteristics of neoplastic and benign lesions. Spectrophotometric traces in every case revealed the absorption bands of oxyhaemoglobin, which led to a consideration of whether the number of red cells per unit volume was the major factor leading to the coloration upon transillumination. The fact that characteristic ultrasound Doppler signals arise from neovascularization around malignant lesions (17,18) reinforced the plausibility of this view. Samples of blood diluted in normal saline were placed in square section cuvettes and placed on an SP8-100 UV-visible spectrophotometer. The traces showed transmission cut-offs over a range of wavelengths. Whole blood was totally absorbing. When the same samples were transilluminated by the Sinus torch and the

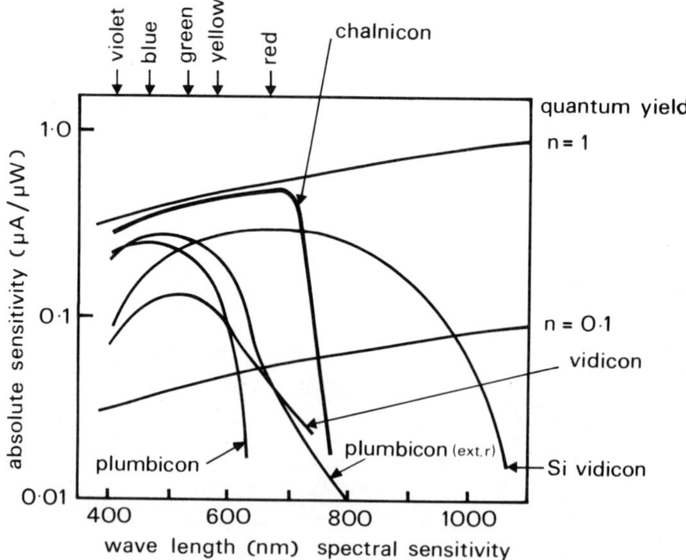

Figure 6.1 The spectral response of various detectors used in television cameras. For tissues transparent in the red and near infrared, chalnicon and silicon vidicon are the most likely candidates for imaging.

images recorded on colour infrared film the colours were in the same range as those of the transilluminated breast. The brightness observed in the case of cysts can be explained by the fact that the tissues containing blood are pushed aside, thus reducing absorption. Malignant lesions with an advancing neovascularized front cause a dark region due to excess absorption. The possibility that some form of *in-vivo* spectrophotometry might further extend the differential diagnostic capability of transillumination has been suggested by Carlsen (19) and by Watmough (15, 16).

6.4 Telediaphanography

The realization by Watmough (15, 16), Mallard (20), Hussey *et al.* (7), Morton and Miller, Watmough (21), Carlsen (19) and Ohlsson (22) of the limitations of diaphanography based on colour infrared film has led to an interest in applying television techniques to producing real-time images of lesions by transillumination. Figure 6.1 shows the spectral response of various detectors. At low light levels the Chalnicon and Silicon vidicon are most suitable. The

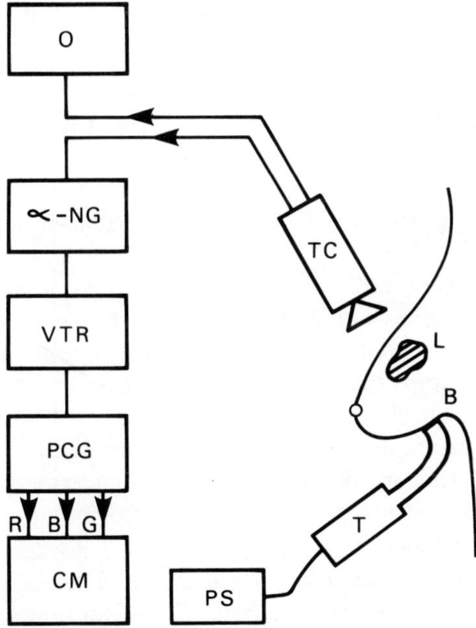

Figure 6.2 A schematic diagram of equipment as used for telediaphanography. T, torch; PS, power supply; B, breast; L, lesion; TC, infrared-sensitive television camera; VTR, video tape recorder; O, oscilloscope; α-NG, alphanumeric generator; PCG, pseudo-colour generator; RBG, red, blue, green inputs; CM, colour monitor. Details of torch are given in reference 15.

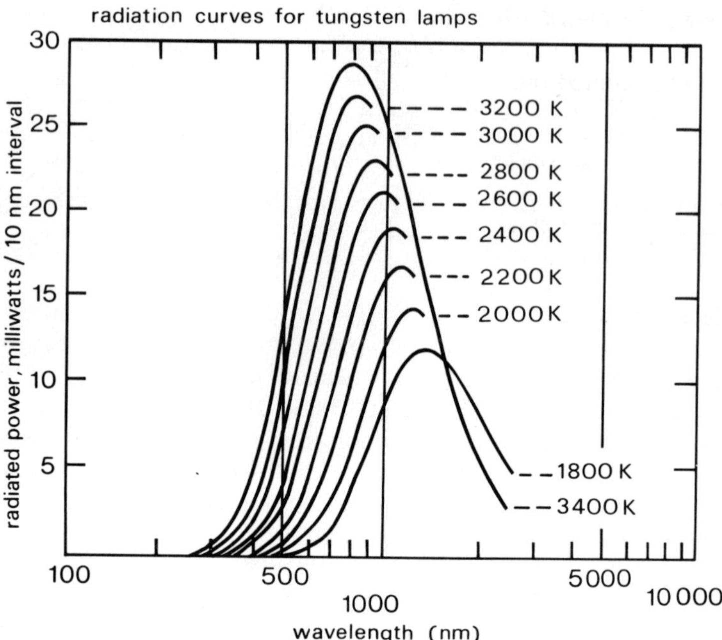

Figure 6.3 Relationship between radiated power from tungsten filament lamp and wavelength. Each curve represents the power distribution for a 1-watt emitter at the indicated colour temperature.

fact that the silicon detector is sensitive out to 1000 nm led to its choice in the author's system. Figure 6.2 shows a schematic arrangement of torch, camera, α-numeric generator, pseudo-colour generator, video recorder and monitor. The detecter is sufficiently sensitive to permit imaging with a tungsten filament lamp without the need for a xenon flash tube. This is advantageous as it avoids the need for high voltages in the torch. The radiated power from a tungsten filament lamp as a function of wavelength depends on filament temperature as indicated in figure 6.3. The television monitor is positioned behind and above the patient. An unexpected benefit with this arrangement is that the torch may be positioned for optimal imaging of a given lesion. The α-numeric generator enables the video recording to be coded with date, patient identification and site. The additional diagnostic value of pseudo-colour images (Plate II, C, D) has still to be assessed; however, from a practical point of view, they do not upset the dark adaption of the observer. There is as yet no published data on the efficacy of telediaphanography though work is under way in Sweden, America, Canada and the United Kingdom.

6.5 Screening for breast disease

Strax (10) has pointed out that

(a) breast cancer is the most common form of cancer in Western women
(b) breast cancer shows an increasing incidence for all age groups
(c) breast cancer is the major cause of mortality in women aged 39-44
(d) there has been a stationary death rate from the disease in the United States for the past 40 years.

These factors were the motivation for the Health Insurance Plan mass screening study (by X-ray mammography and clinical examination) which Strax claims has led to a 25% reduction in death rate after 13 years' follow-up. Strax further argues that the potential value of diaphanoscopy (viewing in darkened room during transillumination) and diaphanography (recording on film) can be delineated as follows:

(1) In symptomatic women where the question of radiation hazard from mammography is moot, diaphanography offers a method of distinguishing a benign lesion, particularly a cyst, from a malignant lesion. It may also be an additional method for detecting a cancer that is questionable on palpation or even mammography.
(2) It may thus make it possible to avoid mammography in many women.
(3) In mass screening, about 10% of cancers occur within a year of a supposedly negative examination (interval cancers). An additional method like diaphanography, which may be able to detect cancer not found by the other modalities, could be useful in adding to the yield of palpation and mammography and thus increase the sensitivity of mass screening.
(4) Finally, if diaphanography turns out to have an accuracy equal to or greater than mammography and if transillumination can detect the same cancers found on X-ray, it may indeed replace the radiological procedure in mass screening. All these potential advantages of diaphanography still remain possibilities at this time (10).

There have so far been only two screening series using transillumination which show some promise. Ohlsson *et al.* (4) reported on 1500 well women who were examined. There were 42 suspected lesions, 10 occult carcinomas of which 4 were not detected by mammography, and 5 where there was no clinical abnormality. More recently, Carlsen (14) refers to 5000 patient examinations on 3293 patients. 1283 women formed an asymptomatic sub-group, who were initially examined by non-ionizing techniques including physical examination, light scanning (transillumination) and, to a lesser extent, ultrasound. 7 patients with breast carcinoma were found in this sub-group, the diameter of the largest lesion being 22 mm, the three smallest being non-palpable. In the overall population, 148 lesions were found, of which 25 were malignant, 21 were positive by X-ray mammography, 18 by ultrasound and 32 by light scanning. The smallest number of false negatives was obtained by light scanning.

In addition to the favourable early indications about the efficacy of diaphanography reported by Ohlsson *et al.* (4), Hussey *et al.* (7), and Carlsen (19), a series of papers by Di Maggio *et al.* (11, 23) showed that in a diagnostic series 96.8% of breast carcinomas were opaque (positive) with transillumination. Di Maggio recommends diaphanoscopy as a technique to be used with thermography in screening for breast carcinoma. In a series of 860 mammary carcinomas Di Maggio was able to compare the results of (a) clinical examination, (b) clinical + mammography, (c) clinical + mammography + thermography, and (d) clinical + mammography + thermography + diaphanography. The percent correct diagnoses were 78.5, 91.7, 96.5 and 98.1 respectively. The evidence available thus appears to show that diaphanography can be a valuable supplement to mammography and clinical examination in the diagnosis of breast lesions.

In relation to screening of the asymptomatic female population for non-palpable breast lesions, a potential method should ideally (1) be capable of imaging non-palpable lesions i.e. much smaller than 10 mm diameter, (2) indicate the nature of the lesion, (3) carry no risk, and (4) be rapid and inexpensive. In the United States, mammography has been discontinued as a screening tool for women in the age range below 50 unless at special risk (24). This is because the risk benefit index for that group was found to be negative from data obtained in the study carried out by the Health Insurance Plan of New York. Doppler studies of the breast when no well-defined site is apparent can be lengthy and require the collection and analysis of a large amount of data. The possibility of utilizing transillumination and ultrasound together to overcome this problem is worthy of consideration.

It is not clear at this stage whether the technique of transillumination can be improved to the point where it might meet the criteria set out above. Certainly it is unlikely to provide resolution comparable with that of X-ray mammography. However, the optical absorption properties of tissues do offer additional information and further efforts to refine the technique seem to be fully justified. Further development of transillumination as an imaging technique seems likely in the near future, as indicated by British patent applications 8104279, 8134648 and GB 2075668. This latter publication also indicates the possibility of spectrophotometry *in vivo* as a method of differential diagnosis.

Acknowledgements
The author wishes to thank the Scottish Home and Health Department for financial support and the suggestion to develop model lesions; the Ross-shire Health Protection Trust who kindly loaned equipment for transillumination, and father (J. W) who always encouraged me to ask questions, silly or otherwise.

References

1. C. H. Cartwright, *J. Opt. Soc. Am.* **20** (1930) 81–84.
2. M. Cutler, *Surg. Gynecol. Obstet.* **48** (1929) 721–729.
3. Ch. M. Gros, Y. Quenneville, Y. Hummel, *J. Radiol. Electron.* **53** (1972) 297.
4. B. Ohlsson, J. Gunderson, and D. Nilsson, *World J. Surg.* **4** (1980) 701–707.
5. H. Bailey, in *Clinical Surgery*, Bristol; John Wright & Sons Ltd. 14th edn., (1967) 155.
6. H. Bailey, Lancet **1** (1936) 902.
7. J. K. Hussey, A. F. MacDonald, D. M. Nicols and D. J. Watmough, *Brit. J. Radiol.*, **54** (1981) 163.
8. M. Baum, *Recent Advances in the study of Breast Cancer*. Cancer Research Compaign Annual Report (1980).
9. R. G. Mole, *Brit. J. Radiol.* **51** (1978) 401–405.
10. P. Strax; *Med. Progr. Technol.* **7** (1980) 131–137; P. Strax, invited commentary, *World J. Surg.* **4** (1980) 701.
11. C. Di Maggio, *Estratto da Radiografia Medica*, No. 41, Kodak (1977).
12. J. D., Hardy, C. Muschenheim, *J. Clin. Invest.* **13** (1934) 817–831.
13. J. D. Hardy, H. T. Hammel and D. Murgatroyd, *J. Appl. Physiol.* **9** (1956) 257–64.
14. P. H. Colins, and W. E. Harper, *Trans. Illum. Eng. Soc.* **20** (3) (1955) 109–119.
15. D. J. Watmough, *Brit. J. Radiol.*, **55** (1982) 142–146.
16. D. J. Watmough, *Acta Radiol Oncol.* **21** (1982) 11–16.
17. P. N. T. Wells, M,. Halliwell, R., Skidmore, A. Webb, and J. P. Woodcock, *Ultrasonics*, **15** (1977) 231.
18. P. N. Burns, M., Halliwell, P. N. T. and A. J. Webb, *Ultrasound in Medicine and Biology* **8** No. 2, (1982) 127–143.
19. E. Carlsen, *Diagnostic Imaging*, April 1982, 28–33.
20. J. R. Mallard, Silvanus Thompson Memorial Lecture. *Brit. Radiol.* **54** (1981) 831–849.
21. R. Morton, & S. S. Miller, *J. Audiovisual Media in Med.* **4** (1981) 86–90.
22. B. S. Ohlsson, (1982), private communication.
23. C. Di Maggio, L., Pescarini, A., Cavallo, R. Duo, and G. Ricci, *Senologia* **5** (1980) 27–32; C. Di Maggio and L. Pescarini, *Senologia* **1** (1980) 26–29.
24. L., Breslow, B., Henderson, F. Massey, *et al. J. Natl. Cancer. Inst.* **59**, No. 2 (1977) 475–477.

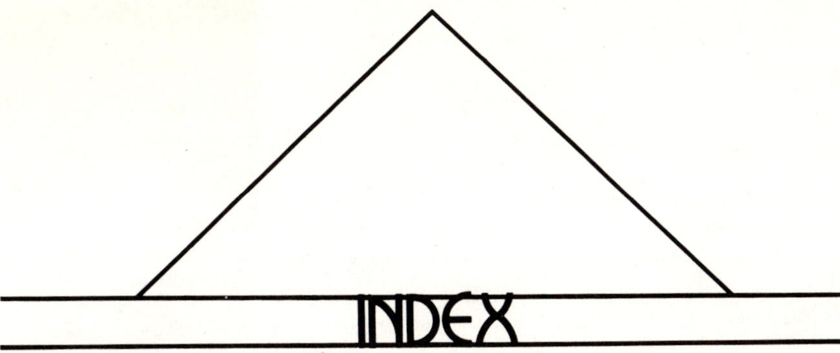